Science of Electricity

Volume 9

Transmission of Electrical Power Explained Simply

by Mark Fennell

© 2012

This book is part of the
Energy Technologies Explained Simply™ Series

Other Books in the Energy Technology Series

Renewable Energy Books
Introduction to Electrical Power
Hydropower Technologies Explained Simply
Wind Power Technologies Explained Simply
Solar Power Technologies Explained Simply

Coal Power Books
Coal Power Technologies Explained Simply
Formation and Mining of Coal
Clean Coal Technologies
Mercury and Coal Power

Nuclear Power Books
Nuclear Power Meltdowns and Explosions
Health Hazards of Radioactive Decay
Radiation Measurements
Processes of Radioactive Decay and Storage of Nuclear Waste

Natural Gas Books
Natural Gas Basics
Extracting and Refining Natural Gas (includes Fracking)
Transportation, Storage, and Use of Natural Gas

Power Line and Grid Books
Introduction to the Transmission of Electrical Power
Power Lines
Underground Cables
Utility Operations and Quality Control
Power Grids Explained Simply

About the Book

<u>Overview</u>

This book discusses everything you need to know regarding the transmission and distribution of electrical power. In this book you will learn the sequence of events and the technologies available for sending electrical power from the power plants to the appliances.

This book is also a "field guide" (with numerous pictures) designed to help you identify components of the transmission system which you may see every day around you.

<u>Importance of a Well-Designed Transmission System</u>

Improperly designed transmission systems can lose significant amounts of power. There are regions of the world with transmission systems so poorly designed that most of the power is lost before reaching the home or business.

Furthermore, the amount of power lost, even for otherwise well designed systems, is directly related to the distance of the power line. While well-meaning decision makers are building longer and longer power lines, we are losing more and more of the power we generated, before the end user sees any of it.

On the other hand, creating the right transmission system can reduce our power loss, which is another way of conserving energy. Therefore, rather than only using a "turn off the lights" approach to saving energy, we can save energy on a much grander scale simply through developing properly designed power transmission systems.

<u>Factors to Consider, Factors to Discuss</u>

There are many factors to consider when creating a transmission system for electrical power. The first level of factors includes placement of power lines, type of transformers, and use of substations. We then must select the material and size for our wires.

We must also consider whether to place lines above ground or below ground, and how to protect the power lines against damage from weather and geology.

As we reach the end user there are other factors to consider, such as what voltage to use, how many phases of power, and many other options related to circuitry.

Therefore this book discusses many factors, all of which will guide the reader to properly designing the best electrical power transmission and distribution system for any situation.

Specific topics covered in this book

The first chapter provides an overview of the electrical power transmission system. You will get a broad perspective. Here you will learn the main components and the sequence of the process.

Chapter two discusses transformers and substations. In this chapter you will learn how a transformer works, the terminology associated with a transformer, and how to calculate voltage change through any transformer. You will also be able to identify the main transformer types and arrangements. The chapter concludes with an overview of substations, focusing on the main possible uses of substations.

Chapter three provides an overview of power lines. Two of the most critical decisions when creating the best transmission system are selecting the right power line and placing the power line in the best location. Therefore this chapter provides an overview of the factors to consider when selecting and placing power lines. Note that most of the rest of the book discusses those factors in detail.

Chapter four discusses high voltage transmission lines. In this chapter you will learn the types of high voltage lines. You will be able to identify possible arrangements of high voltage lines on transmission towers. You will learn the best choices for materials for the power lines. A separate section is devoted to weather tips: how to ensure that power lines survive the most extreme weather. The chapter ends with a detailed discussion of safety of high voltage lines, including the effects of EMF on human health.

Chapter five discusses the lower level voltage lines which are used in neighborhoods. In this chapter you will learn about the design and maintenance options for the power lines near your home.

Chapter six is devoted to the new technology of high voltage direct current (HVDC) power lines. Used properly these lines are beneficial, yet used improperly these power lines will waste enormous amounts of power. In this chapter you will learn the advantages, disadvantages, and best uses of HVDC power lines.

Chapters seven and eight discuss underground cables for power lines. Underground cables allow the region to look neater, yet there are many difficulties associated with underground cables. The primary concerns are the effects of geology and water on the power lines. Other issues include limited access for maintenance, increased cost, and interference with other underground structures.

Therefore in chapter seven you will learn the advantages and disadvantages for underground placement of power cables.

In chapter eight you will learn some of the most important practical tips for installing underground cables (primarily to protect against geological effects).

The final chapter discusses the sequence of power through the homes and businesses. The topics in this chapter relate to the components and events in your home. Here you will learn exactly how the electrical power flows from the transformer outside your home, through your home into the appliance, and outward again. You will also learn some basic home electrical concepts, such as prongs and outlets, variation in voltages, and colors of wires.

In this chapter you will also learn how businesses use electrical power for their needs. You will learn the design options which should be considered when creating an electrical distribution system for large facilities.

At the end of the book you will find an Appendix with two data tables: 1) wire sizes in different units, and 2) resistance in wires based on type of material and wire size.

In total, everything in this book will help you design the best power transmission system for your region and needs. This book will also be your field guide, helping you understand the components of electrical power distribution that you see around you every day.

About the Energy Technology Series

The books in the *Energy Technologies* series are designed to educate citizens, students, and legislators on all aspects of energy technologies. The first books in the series focus on electrical power.

The books discuss many energy technologies, including: generators, turbines, power plants, power lines, and grids. The technologies for each type of power source (hydro, wind, solar, coal, nuclear, and natural gas) are discussed in detail. The books also discuss efficiency, safety, reliability, and health concerns for each energy technology.

The ultimate goal of the series is to enable the people to make informed decisions on practical energy questions. The secondary goal is to serve as introductory guides for students embarking on careers with energy technologies.

Taken altogether, the books in the series answer any question you are likely to have, such as:

- How can we increase the efficiency of solar cells?
- How do I select the size my solar array?
- What do I need to know when installing a wind turbine?
- How effective are the clean coal technologies?
- How can we prevent grid failures?
- Do power lines cause cancer?
- and many other energy technology questions...

Science of Electricity in Perspective

The subject of electrical power is of great importance to our communities, but is rarely taught. Public debate is frequent and passionate, but with too little understanding of the actual science. At best, an informed citizen knows only a few pieces. At worst, as it is for a great number of citizens, electricity is magic and myths are believed as scientific truth. It does not have to be that way. Any citizen, regardless of background, can know the technologies behind all aspects of electricity.

The books in this series solve that problem. These books educate the general public in all aspects of electrical power. Any person, regardless of background, can easily find the answer to his energy question in one of these books.

Specific Goals

There are numerous technologies described in these books. Yet for each technology I sought out the answers to the following questions:

1. How does the technology work?
2. What are the advantages and disadvantages?
3. What is the efficiency? How can the efficiency be improved?
4. What is the environmental impact? How can it be improved?
5. What are the safety hazards, and how can they be reduced?
6. What are the most important practical tips?
7. What facts comprise the most important data?

Technical Discussions Explained Simply

The books in the series must necessarily be technical to some degree. Electricity is a practical technology, and therefore we must understand the technical aspects if we want to make wise decisions. Yet the discussions in this book are always aimed at the citizen or policy maker.

The books in this series explain the principles of electricity as simply as possible, using ordinary English (no engineering jargon), and highlighting the most important points of each technology. Main concepts and facts are emphasized with the use of lists, tables, diagrams, and summaries.

I do not expect any reader to have a background in science, yet I offer enough facts and details so that the reader can have an accurate understanding of all related technologies. I provide enough technical details and enough data for the reader to make informed decisions.

Conclusion

For all the reasons above, I offer this series of books. My goal is to inform you on the basic concepts of all the technologies and all of the issues related to electricity so that you can make realistic decisions.

Remember that there are no perfect solutions, there are only choices. I hope that this series of books will assist you in making those choices for your community.

Mark Fennell

Accuracy and Technical Depth

Objectivity

I have tried my best to be as objective as possible. Whereas many other authors of energy books have an agenda, I have no desire to promote one industry over another. I have no desire to promote one technical solution over another. In this endeavor, I have tried to be an objective scientist.

Accuracy of Data and Summaries

I never relied solely on the conclusions of other researchers. Instead, I performed many other tasks to ensure that all conclusions were accurate. I examined primary data whenever possible. I have read the fine print on how research was obtained.

I have also checked the accuracy of the conclusions written by other researchers, most commonly by finding at least three distinct sources for each fact. In addition, I performed my own calculations numerous times to prove (or disprove) conclusions and final values in other reports. It is only after such rigorous investigations that I created data tables and wrote summaries for these books.

Limited Mathematics

The books must also use math from time to time. For example, efficiency is a statement of a specific amount, and therefore the discussion of efficiency requires the use of equations. Other issues such as power loss, health hazards, environmental concerns, and quality control are also statements of amounts and also require calculations. Therefore some equations are necessary to know, even for the non-scientist.

I also provide examples of calculations so that readers can become more comfortable with using the equation themselves.

However, I want to emphasize that I focus on concepts not on the mathematics. I provide equations only when it is necessary for the citizen or student to be familiar with these equations.

Table of Contents

Selected List of Figures

Table of Contents: Detailed

9.1
The Sequence of Transmitting Electricity from a Power Plant to Your Home

Introduction

In this chapter we will provide a brief overview of the main components of the transmission system: transformers, substations, and power lines. We will then present the basic sequence of events which sends electricity from a power plant to your home.

List of Topics for this Chapter
1. Transformers, Substations, and Power Lines Overview
2. Sequence of Electricity from a Power Plant to your Home
3. Adding Sub-transmission Substation and Other Power Plants

Overview of Transformers, Substations, and Power Lines

Introduction

In order to get electricity from the power plant to the homes and businesses, we need three types of equipment: transformers, substations, and power lines.

Transformers

Transformers exist to change the voltage. We could not have the electric power that we have come to rely on without transformers. This is due to two factors: inherent power loss in transmitting electricity through wires, and size of the wires. Changing the voltage helps for both those issues. Specifically: a) higher voltage reduces power loss, and b) higher voltage also means that smaller diameter wires are needed. The net result is that we can transmit power over hundreds of miles, with a reasonable efficiency.

There are at least two transformers in any power system, and often several transformers. We need one transformer to increase the voltage so that we can transmit the electricity with minimal power loss.

Yet we also need another transformer to reduce the voltage to safe levels in the communities. Transformers come in many sizes, from the large ones at substations to the smaller ones on poles.

Substations

The main purposes of a substation are: as the location for a transformer, to start distributing the power through the community, and to link power coming in from various power plants. Usually several substations are used in sequence, from power plant to a home. In addition, larger cities usually have several substations, one substation located in each neighborhood.

Power Lines

We need wires to carry the electricity. Therefore, we have a series of different wires and cables, each carrying differing voltages. There are three categories of power lines: High Voltage Transmission Lines, Community Level Distribution Lines, and Underground Cables. Each type of power line has its own technological issues.

Sequence of Electricity
From a Power Plant to Your Home

Introduction

In this section we will present the basic sequence of events to get electricity from a power plant to your home. As we do this, you will see where each transformer, each power line, and each substation fit into the sequence. (Figure 9.1)

List of Steps in the Sequence (Figure 9.1):

1. Power Plant
2. Substation #1, which is also Transformer #1
3. High Voltage Transmission Lines
4. Substation #2, which also has Transformer #2
5. Primaries
6. Distribution Substation (Substation #3)
7. Secondary Wires
8. Final Transformer (pole or pad transformer)
9. Service Wires

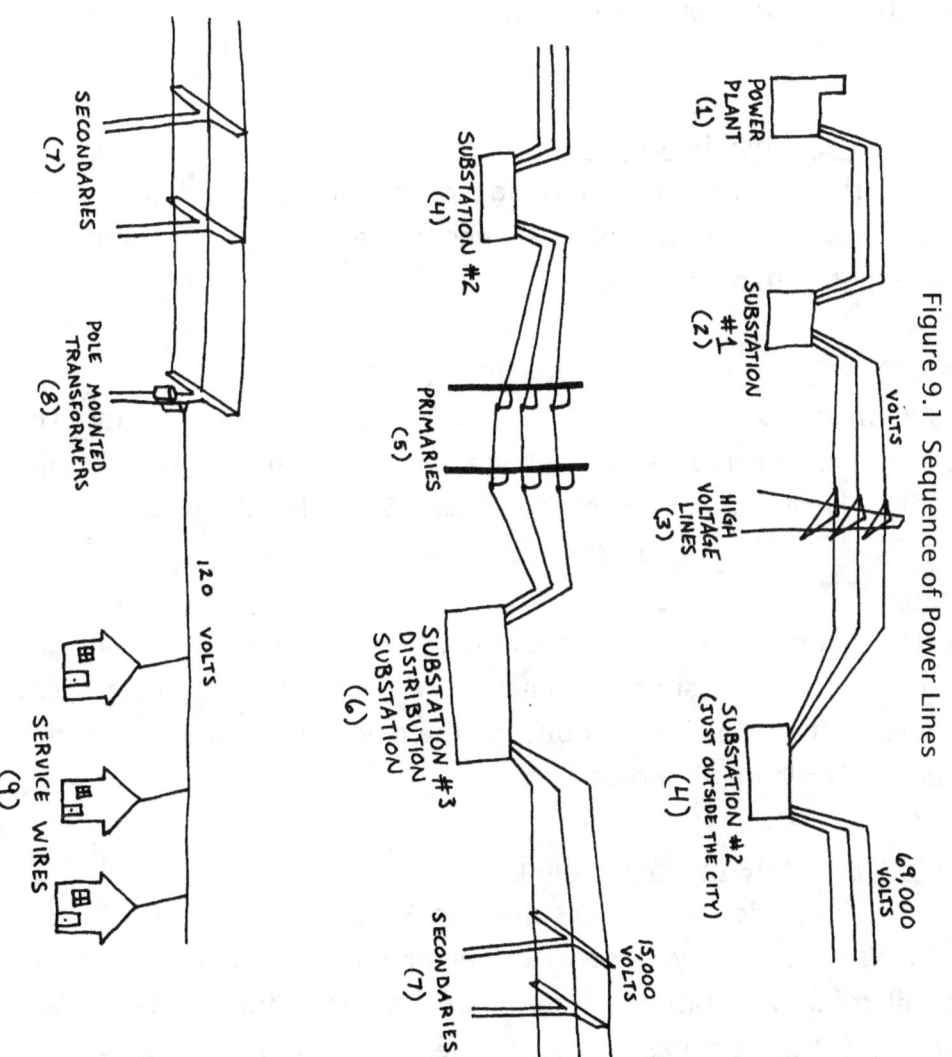

Figure 9.1 Sequence of Power Lines

POWER PLANT (1)

SUBSTATION #1 (2)

HIGH VOLTAGE LINES (3)

SUBSTATION #2 (JUST OUTSIDE THE CITY) (4)

69,000 VOLTS

VOLTS

SUBSTATION #2 (4)

PRIMARIES (5)

SUBSTATION #3 DISTRIBUTION SUBSTATION (6)

15,000 VOLTS

SECONDARIES (7)

SECONDARIES (7)

POLE MOUNTED TRANSFORMERS (8)

120 VOLTS

SERVICE WIRES (9)

1. Power Plant
The power plant generates electricity.

2. Substation #1, which is also Transformer #1
Substation #1 is located very close to the power plant. A transformer at this first substation increases the voltage to a very high amounts, such as 345,000 Volts or more, so that we can transmit power with minimal loss.

3. High Voltage Transmission Lines
The electricity is carried across dozens of miles, even hundreds of miles, over high voltage transmission lines to the cities. Tall metal towers hold these high voltage lines.

4. Substation #2, which also has Transformer #2
Substation #2 is located just outside the borders of the city. The transformer here decreases the voltage to a level that is safe for the community. The voltage is lowered to at least 69,000 Volts.

5. Primaries
The first power lines that carry voltage less than 69,000 Volts are called primary wires, or simply "primaries." Primary wires carry electricity from the substation which is just outside the city limits to the distribution substations in each neighborhood.

6. Distribution Substation (Substation #3)
The next substation is the Distribution Substation. The Distribution Substation is located very close to the neighborhood. There are usually several distribution stations, often just a few miles apart. Each Distribution Substation has a transformer which reduces the voltage another step, such as down to 15,000 Volts. The situation is similar to the post office. Just as letters are sent to a central post office in each neighborhood, so electricity is sent to distribution substations in various regions of the city.

7. Secondary Wires

Secondary wires, or "secondaries", are wires that carry electricity from each Distribution Substation into the local neighborhoods. Secondaries are the electrical wires that you see lining poles along the street. These wires can also be underground cables, and if so, these wires also follow the paths of the streets. A typical voltage for secondaries may be 15,000 Volts.

8. Final Transformer: Pole Mounted Transformer or Pad Transformer

The final transformer is a smaller transformer, usually attached to the side of an electrical pole on the street. No longer do we need a substation to house the transformer, a small cylinder on a pole will do. The distribution transformer takes the voltage down to its final use for homes, typically 120 Volts.

9. Service Wires

The final step is the service wires. These wires carry the final home use voltage, usually 120 Volts, directly to the homes.

Adding a Sub-transmission Substation and Other Power Plants to the Sequence

For larger cities, we need to add equipment to the sequence. Larger cities will often have these additions: 1) power coming in from other nearby power plants, and 2) multiple distribution substations. Both of these additions require the Sub-transmission Substation.

Note that the sub-transmission substation does *not* usually change voltages. This substation is linked to the various distribution substations throughout the city.

The sub-transmission substation can also work with other power plants. When power plants are linked on a "grid", the actual lines are usually linked at the sub-transmission substations.

Chapter Summary

1. Transformers exist to change the voltage.

2. We need transformers in order to increase the voltage so that we can carry electric power over long distances. However, large voltages are dangerous. Therefore the voltages must be reduced before reaching the customers.

3. There are three categories of power lines: High Voltage Transmission Lines, Community Level Distribution Lines, and Underground Cables. Each type of power line has its own technical issues.

4. The main purposes of a substation are:
    a. as the location for a transformer
    b. to start distributing the power through the community
    c. to link power coming in from various power plants

5. The general sequence from power plant to the home is as follows:

    After the power plant generates the electricity, a transformer at a substation near the power plant increases the voltages to very high voltages. The electricity is then carried across hundreds of miles via high voltage transmission lines.

    Near the city, at a second substation, a second transformer reduces the voltages from high voltages to community levels. Wires carry electricity from this transformer & substation to distribution substations in each neighborhood.

    At the distribution substation in your neighborhood a transformer reduces voltage to yet another level. Electricity is carried from the local distribution substation to the final transformer (near your house). These wires may also be laid underground beneath the streets.

    A few blocks from your home, a final transformer (a small transformer, usually on a pole) reduces the voltage to its final level of 120 Volts. The final wires, called service wires, carry the 120 Volt electricity directly to each home.

9.2
Transformers and Substations

Introduction

We could not have the electric power that we have come to rely on without transformers and substations. Transformers exist to change the voltage. This is necessary in order to reduce power loss. Substations are primarily used as locations for transformers. However, a substation can have many other uses.

In this chapter we will explain how transformers operate, and we will describe the main types of substations.

List of Topics for this Chapter
1. Overview of Transformers
2. Power Loss and Transformers
3. Wire Size and Transformers
4. Main Terms Regarding Transformers
5. Coil Windings and Voltages
6. Transformer Arrangements
7. Linking Transformers Together
8. Multiple Transformers and Safety
9. Possible Uses of a Substation
10. Types of Substations

Overview of Transformers

Transformers exist to change the voltage. Each transformer is designed specifically to either increase the voltage by a specific amount, or decrease the voltage by a specific amount.

A transformer has two coils of wire. The first coil carries the electricity that is coming in. The second coil carries the electricity that is going out. Electricity runs through the first coil, which then induces a current in the second coil. A change in voltage will occur if the second coil has a different number of windings than the first coil.

There are always two transformers in a power system. The first transformer exists near the plant. Its purpose is to increase the voltage so that it can be transmitted over long distances.

The second transformer exists near homes. Its purpose is to decrease the voltage so that it can be used safely by the people. Usually the decrease in voltage occurs in steps through several transformers.

Transformers come in many sizes, from the large ones at substations to the smaller ones on poles.

Power Loss and Transformers

As electricity is transmitted across wires, power is lost. This power loss is due to the resistance in the wires.

There are two important concepts to remember regarding power loss on wires:

1. Power loss will happen with any wire, even to some degree in the most conductive materials (including copper and aluminum).

2. A longer power line results in greater total power loss. Therefore any power line, even using the best conductors, will eventually lose most of the power.

Without a transformer most of the power generated at any plant would be lost before reaching the homes. Transformers simultaneously increase the voltage and decrease the current, while keeping the amount of power constant. The net result is much less power is lost during transmission. We can therefore transmit electricity over longer distances, with reasonable efficiency.

Recall the equation: Power Loss = (Current)^2 x Resistance. Looking at this equation, we can see that we could reduce the power loss in two ways: 1) by reducing the current or 2) reducing the resistance. First we reduce the power loss by focusing on the resistance. We get the lowest resistance by selecting the most conductive material for our power lines. The best materials include copper and aluminum.

After the wires are in place, the only way to reduce the power loss further is to reduce the current. This is done by increasing the voltage. Recall that Power = Voltage x Current. Therefore, if we increase the voltage, and if we intend on transmitting the same amount power, then the current must naturally decrease. Then, because we have decreased the current, the power loss will be reduced.

Wire Size and Transformers

Power loss is the primary reason why power cannot be transmitted long distances without transformers. In addition, there is another factor: the size of the wires. The size of the wire that we need depends directly on the amount of current that we expect to put through that wire. In order to carry more current, we need larger wires. However, the larger the wire, the heavier it becomes. Therefore, wire size becomes a practical issue.

Recall that we increased voltage in order to reduce power loss. Also recall that when we increase the voltage, the current must decrease. We then apply this fact to wire size: if the current is less, then we need a smaller wire. This becomes a benefit when we start lining wires over long distances.

Main Terms Regarding the Workings of Transformers

The following are the main terms you will see associated with the technology of transformers:

1. Primary coil, Primary winding, or Primary: The "primary" is the coil that carries the voltage which enters the transformer.

2. Secondary coil, Secondary winding, or Secondary: The "secondary" is the coil which carries the voltage that leaves the transformer. The secondary always carries the changed amount of voltage, whether increased or decreased.

3. Step-up Transformer: A "step-up transformer" is a transformer which increases the voltage.

4. Step-down Transformer: A "step-down transformer" is a transformer which decreases the voltage.

5. Number of windings in the primary coil, abbreviated N_p: This is the number of windings, or number of turns, in the primary coil.

6. Number of windings in the secondary coil, abbreviated N_s: This is the number of windings, or number of turns, in the secondary coil.

7. Voltage in primary coil, abbreviated V_p: This is the voltage that is put into the transformer.

8. Voltage in secondary coil, abbreviated V_s: The voltage in the secondary coil is the new amount of voltage, the amount which leaves the transformer.

Coil Windings and Voltages

Introduction

The amount that the voltage will increase (or decrease) is directly related to the number of windings of each coil. A simple fraction can be set up, which makes calculations easy to do:

$$\frac{\text{Number windings in primary coil}}{\text{Number windings in secondary coil}} = \frac{\text{Voltage in primary coil}}{\text{Voltage in secondary coil}}$$

This relationship can be abbreviated to: $N_p/N_s = V_p/V_s$

Example #1

The number of windings in the primary coil is 10, and the number of windings in the secondary coil is 40. $N_p/N_s = 10/40$, or 1 to 4.

First, we notice that the secondary coil has more windings than the primary. Therefore, this is a "step-up" transformer, where the voltage is increased. Second, no matter what amount of voltage we start with, the secondary voltage in this transformer will be 4 times greater than the voltage we started with. For example, if we put 6,000 Volts into the transformer, then we get 24,000 Volts out.

Example #2

The number of windings in the primary coil is 20, and the number of windings in the secondary coil is 10. $N_p/N_s = 20/10$, or 2 to 1. In this case, the secondary coil has less windings than the primary, and is therefore a step-down transformer. For example, if we put 6,000 Volts in, then we will get 3,000 Volts out.

Transformer Arrangements

Transformer Arrangements: Overview

There are several options for transformer arrangements. The most common options are:

- single phase transformer
- bank of transformers
- single phase transformer with three wires
- three phase transformer

Each of these will be described briefly below.

Single Phase Transformer (Figure 9.2)

Most transformers at any stage of the sequence are "single phase transformers." Recall that three phases of electrical power are produced, and these three phases are transmitted over three distinct lines. Each phase of electricity will need to be transformed in voltage at various points. Therefore each phase of electricity must go through its own "single phase transformer".

Fig. 9.2 Single Phase, Pole Mounted Transformer

Bank of Transformers (Figure 9.3, Figure 9.4)

A bank of transformers is a set of transformers at one location. You will often see three boxes close together. Each box is a single phase transformer. Together, they form a "bank" of transformers.

Figure 9.3 A bank of three, single phase, pole mounted transformers

Figure 9.4 A bank of three, single phase transformers at a substation

For every location where the voltage must be changed, there are usually three transformers next to each other. Each of these three transformers takes care of its own phase. Each of these three transformers acts independently of the others. Therefore, many locations have three of these "single phase transformers" sitting together. This set of transformers constitutes a "bank" of transformers.

Single Phase Transformer with Three Wires (Figure 9.5)

This the most common type of transformer which is found close to the homes. As the three phases get very close to their destinations, these three phases split up. Each phase goes to a different block of homes. At this point, we have the final step-down transformer. This box is just one single phase transformer, with three wires connected to the home. All three wires carry the same phase of current.

Figure 9.5 Single phase step-down transformer with three wires

Let me repeat for clarity: all three wires of the final transformer carry the same phase of current.

These three wires are connected from the outside into a particular home. Two wires are "hot" wires. These wires carry current into the home. The third wire is the "ground" wire. This wire carries current leaving the home. (Details of the pole mounted transformer and its three service wires will be discussed in the chapter called "To the Home.")

Three Phase Transformer

This is essentially all three of the single phase transformers in one container. (It is a bit more complex than that, but that is the basic principle). As with the banks of transformers above, we are still transforming the voltage of each of the three phases independently. However, the entire situation is now in one box rather than in three boxes.

Multiple Transformers and Safety

Using a greater amount of voltage in our power lines will result in greater efficiency. This means that we lose less power and that we can transmit electricity over longer distances. However, we cannot allow those high voltages in our homes or along the streets because those high voltages are quite dangerous. Therefore, the voltages must be reduced before the power reaches the people.

Because the utility company wants to balance efficiency and safety, the result is usually several transformers in sequence. The power lines will carry the highest voltage possible (in order to reduce power loss), while still being safe for the particular area. The power lines will carry this level of voltage for the longest distance possible. When the power lines enter a more populated region, the voltage will be reduced to a safer level.

By the time the wires reach your home, after the final step-down transformer, the voltage is significantly low. (Yes you can still be shocked, yet the result of 120 volts in your body is nothing compared to a shock 700,000 volts!)

Note that although the power lines to your home are far safer than all the other power lines in the sequence, these lines are the least efficient.

Possible Uses of a Substation

Substations are primarily used as locations for transformers. However, there are many other uses for a substation. Most of these other purposes will be better understood in later chapters when we discuss utility operation and quality control. Nevertheless, at this time we will cite the major purposes of a substation. Possible uses of a substation include:

1. Change voltage, using transformers. (Figure 9.4)
2. Connect with other power plants (buying or selling power).
3. Switch circuits into or out of the system.
4. Measure parameters of electric power for quality control.
5. Monitor circuits and communicate with devices remotely.
6. Convert AC to DC.
7. Change the frequency of the current (to help with reliability).
8. Regulate voltage to compensate for system voltage changes.

Types of Substations

Introduction

For practical purposes to most citizens, there are four types of substations:

1. Transmission substations
2. Sub-transmission substation
3. Distribution substations
4. Poles

1. Transmission Substations

The transmission substations work with the high voltage lines. You can remember this because the term "transmission" is most accurately used in reference to the high voltage lines (70,000 Volts and above).

There are always two transmission substations. The first is near the power plant. This transmission substation has a large transformer which increases the voltage to great amounts. The second transmission substation is near the city, and reduces the high voltages to levels acceptable for the community. Power lines which leave this second transmission substation are known as primary wires.

2. Sub-transmission Substation

The sub-transmission substation does not change the voltage. This substation is used for a variety of other purposes, including getting power from other power plants and a variety of quality control issues.

3. Distribution Substations (Figure 9.6)

Distribution substations are located closer to neighborhoods. You will often find these substations along major business roads and small highways. Distribution substations have large transformers, and the main purpose of these substations is to reduce the voltage a second time. Wires that leave here are known as secondary wires.

Figure 9.6 Distribution Substation

4. Poles

When we get to the home, we no longer need a substation to contain a large transformer. At this point, we just have a small transformer, and therefore a pole will be sufficient. In a sense, this pole near your home could be considered the final substation, with the final transformer being on the pole. (Technically, this is not a substation, but I wanted to emphasize the sequence of events.)

Chapter Summary

1. Transformers exist to change the voltage.

2. We need transformers to carry electric power over long distances.

3. Transformers increase the voltage in order to reduce power loss and so that we can use smaller diameter wires. It is only by reducing the power loss and only by using smaller wires that we are able to carry power efficiently over very long distances.

4. However, large voltages are dangerous and therefore the voltages must be reduced before reaching the customers.

5. There are at least two transformers in the sequence, and usually more. The first transformer increases the voltage to a very high amount. The later transformers decrease the voltage, in steps, to the final voltage for the point of use.

6. The step-up transformer increases the voltage. The step-down transformer decreases the voltage.

7. A transformer has two coils of wire. The first coil carries the electricity that is coming in. The second coil carries the electricity that is going out.

8. The first coil is called the primary and the second coil is called the secondary.

9. If the second coil has more windings than the first, then the voltage is increased. If the second coil has less windings than the first, then the voltage is decreased.

10. A simple ratio can be set up, where the ratio of windings in each coil is the same ratio as the voltages of each coil: $N_p/N_s = V_p/V_s$

11. Each phase of current must go through its own transformer. Any transformer which changes the voltage of one of the phases is called a "single-phase transformer."

12. A bank of transformers is a collection of three, single-phase transformers. Each transformer works independently of the others. Each transformer works with just one of the three phases of electricity.

13. A three-phase transformer is a bank of three transformers, where all three are housed in one box. As above, there are three single-phase transformers, each operating independently, only now they are all inside one box.

14. Substations may be used for any of the following purposes:
   a. Change voltage (using transformers).
   b. Connect with other power plants (buying or selling power).
   c. Switch circuits into or out of the system.
   d. Measure parameters of electric power for quality control.
   e. Monitor circuits and communicate with devices remotely.
   f. Convert AC to DC.
   g. Change the frequency of the current (to help with reliability).
   h. Regulate voltage to compensate for system voltage changes.

15. For practical purposes there are four types of substations:
   a. Transmission substations
   b. Sub-transmission substation
   c. Distribution substations
   d. Pole near the home

16. The transmission substations work with the high voltage power lines (70,000 Volts and above). There are always two transmission substations: one near the power plant which increases the voltage, and a second substation near the city which decreases the voltage. Wires that leave the second substation are known as primary wires.

17. The sub-transmission substation does not change the voltage. The main purposes include: linking to other power plants and maintaining quality control.

18. Distribution substations are located closer to neighborhoods, usually along major roads. The main purpose of these substations is to reduce the voltage the second time. Wires that leave here are known as secondary wires.

19. The pole is not a substation in the strict sense, but it is similar. This pole near your home could be considered the final substation, with the final transformer located on the pole.

9.3
Power Lines Overview

Introduction

Power lines are vital to the transmission of electrical power. In order to get electricity from the power plant to the homes, electricity must travel through a series of power lines. The types of power lines can be organized into three broad categories:

1. High Voltage Transmission Lines (power plant to the city)
2. Community Level Distribution Lines (city edge to the homes)
3. Underground Cables

Each of type of power line has its own technical issues. However, all power lines have certain factors in common. In this chapter we will look at the issues common to all power lines. Future chapters will examine the specific issues related to the specific types of power lines.

List of Topics for this Chapter
1. Right-of-Way
2. Temperature and Power Lines
3. Choosing Which Wires to Use: Overview
4. Choosing Which Voltage Wires to Use
5. Choosing Which Material to Use
6. Choosing Which Size of Wires to Use
7. Understanding Wire Sizes
8. Choosing Stranding or Solid
9. Choosing Insulation

Right-of-Way

"Right-of-Way" refers to the area used by the utility company for their towers, poles, lines, and cables. Utility companies must purchase or lease the property. Utility companies also pay significant expense for installing towers and laying cables. Therefore, companies plan carefully before choosing what sites to use. Tips for Right of Way of power lines include:

1. Place power lines along (or below) streets, highways, and railroad tracks.

2. Choose the shortest paths possible.

3. Avoid crossing hills, ridges, and swamps

4. Coordinate with Communication Lines.

    a. lay the wires underground

    b. use the same poles or conduits for both communication lines and electrical lines.

5. Preserve the environment and scenic locations.

6. Designate the right-of-way to be wide enough

    a. A tower needs a right-of-way width that is 50-250 feet (depending on the type of tower).

    b. An electrical pole should have 15-50 feet of width (depending on the type of line and environmental circumstances).

Temperature and Power Lines

Temperature is a critical factor in operating power lines. An increase in temperature can decrease efficiency, damage the power lines, and start fires. Here are a few of the important principles regarding temperature of power lines:

1. A wire will expand as it gets hot and contracts as it cools. If the wire expands and contracts too much, then the wire will eventually break. Therefore, the material of any power line must be able to expand and contract accordingly.

2. When electric current flows in a wire, the wire will become hot. If a wire becomes too hot then that wire can become damaged, even permanently destroyed. Therefore, the power line must either be cooled or be designed to withstand high temperatures.

3. Air naturally cools the power lines which exist above ground. Usually, the natural circulation of air is enough to cool the wires.

4. The temperature of the air around the transmission lines limits the power that the wire can carry. If the air around the wire is cooler, then more power can be transmitted. If the air around the wire is hotter, then less power can be transmitted.

5. There is no natural cooling in underground cables. Therefore, the materials must be designed to withstand higher temperatures, or the power lines must be cooled artificially.

Choosing Which Wires to Use: Main Factors

It is important that we choose the correct wire for each section of power line. There are two reasons for why proper wire selection is so important: 1) potential for power loss along the line, and 2) potential for damage to the line from the natural world.

Choosing the exact configuration for a power line is not simple. There are several factors in choosing which wires to use. The main factors are:

1. Choosing Which Voltage will be Required
2. Choosing Which Material to Use
3. Choosing Which Size of Wires to Use
4. Choosing Stranding or Solid
5. Choosing Insulation (how much, what types)

Note that the topic of selecting wires is important enough that the remainder of this chapter will be devoted to this topic. In the remainder of this chapter we will look at each of the factors listed above.

Choosing Which Voltage to Use

The first step in choosing the wire is knowing what voltage the wire will carry. Choosing which voltage to use is not simple, nor is it entirely scientific. Usually several voltages are considered, and the final decision is often subjective. The following are some important points regarding voltage selection:

a. Operating costs: Higher voltage lines cost more, due to the types of the wires and size of the towers.

b. Allowable power loss: The voltage should be high enough so that the power loss is the maximum allowable.

c. Voltage of existing power lines: Existing power lines or existing substations often dictate the voltage of new lines.

Choosing Which Material to Use

Introduction

There are really only three basic materials used for power lines: copper, aluminum, and steel. However, the properties of these materials differ, and no material is perfect. Therefore, the three basic materials are often used together in various hybrid forms.

There are several factors to consider when choosing the material for the wire: 1) Conductivity, 2) Rust Resistance, 3) Weight, and 4) Strength. In this section we will look at each of these factors.

Conductivity

The three most commonly used metals for conductors are copper, aluminum, and steel. Of the three materials, copper is the most conductive. There are a few metals more conductive than copper but these are too expensive to put into wires.

Copper sets the standard for electrical conductivity in wires. Aluminum is second in conductivity, and is considered a good conductor. However, aluminum is only about 2/3 (specifically 63%) as conductive as copper. Steel is the least conductive of the three main materials. Steel is only about 15% as conductive as copper.

Rust Resistance/Corrosion Resistance

Aluminum is very resistant to rusting, and can last many, many years. Copper is second, also lasting many years. Steel rusts very easily, lasting only a few years before rusting beyond use. This is one of the reasons why steel conductors are given a coating of Copper or Aluminum. The coating on the steel prevents the steel from rusting.

A brief mention should be made of galvanized steel. Technically, galvanized steel is steel coated with zinc. In this respect galvanized steel is similar to the coated steels mentioned above. However, galvanized steel is not useful for power lines because zinc is a relatively poor conductor. Zinc is only about 1/4 as conductive as copper, which is not much better conductivity than steel by itself.

Weight

Using a light weight material is a good choice for power lines which are hung above ground. However, note that light-weight materials do not usually have much strength. Aluminum is the lightest of the conductors, weighing only 33% the weight of copper. Copper is second in weight. Steel is much heavier than the other two.

Strength

The strength of the wire can be an important factor, particularly in areas of high winds, snowstorms, and ice. Where nature adds such extra weight to the wires, the wires must be strong enough to not break.

Aluminum has the least strength: the strength of the aluminum wire is about 50% the strength of the copper wire. Steel is by far the strongest of the conductors. This is why most high voltage power lines have a core that is made of steel.

General Points on Materials for Wires

Copper is a great electrical conductor: it is cheap, it is lightweight, and it resists corrosion well. However, copper has very little strength.

Aluminum is also a good conductor (second to copper). Aluminum is cheap, lightweight, and is lightest of all the conductive materials. Aluminum is also the most rust-resistant material. However aluminum (like copper) does not have much strength.

Steel is a very strong wire. However, steel is not as good a conductor as copper or aluminum. For example, steel is only 15% as conductive as copper. Steel also rusts easily.

Each material is used in different situations. In addition, there are hybrid materials which use the best traits of each individual material.

Hybrid Materials

The most common hybrid materials are ACSR, Copperweld Steel Conductor, and Alumoweld Steel Conductor. Each is designed with the same basic purpose: the steel provides strength, while the copper or aluminum provides corrosion resistance and better conductivity.

1. ACSR (Aluminum–Conductor Steel Reinforced)

The ASCR is several strands of conductors woven together. The center strand is steel for strength. All of the surrounding strands are aluminum (Aluminum is chosen for corrosion resistance, lighter weight, and better conductivity).

2. Copperweld Steel Conductor

This material is a steel wire, with a copper coating. The steel provides the strength. The copper coating provides the better conductivity.

3. Alumoweld Steel Conductor

This is a steel wire, with an aluminum coating. The steel provides the strength. The aluminum coating provides corrosion resistance and better conductivity.

Electrical Insulation Around the Wire

Introduction

The purpose of electrical insulation is to reduce power loss. Without enough electrical insulation the electricity that we have created will flow off the wire to the environment, and not reach any consumer. If we have adequate electrical insulation around a wire then we will prevent electricity from leaving that wire, and more electricity will reach the homes and businesses.

Air as Insulator

Air is a great insulator. Air does not usually conduct electricity, and it is available around all overhead wires. For most situations, air is sufficient as an insulator. However, sometimes air is not enough. The air molecules can become charged. This occurs either through natural weather or through corona discharge. (See the chapter on High Voltage Transmission Lines for more details).

Insulation for Underground Power Lines

Proper insulation is most important in underground power lines. In underground power lines there is no air, and hence no natural insulation as in the above-ground lines.

Furthermore, many soils are electrically conductive, and electrical power would easily drain off the lines if the lines were not insulated.

Therefore in underground power lines we must wrap insulating material around the wire. There are many good materials for insulation including rubber, thermoplastics, and taped insulation.

Wire Sizes of Power Lines

Introduction

Choosing which size of wire to use is not simple. There are several factors to consider. These factors include: 1) Line Voltage; 2) Amount of Power to be Transmitted; 3) Acceptable Power Loss; and 4) Mechanical Strength.

Note that for the first three factors we calculate the amount of current, which will then determine the diameter of the wire. (See volume 1 for details). How these factors affect wire size can be summarized as follows:

1. Line Voltage: for Minimum Wire Size

From the planned line voltage we can calculate the minimum wire size needed. (Current = Power/Voltage. See Volume 1 for details.)

2. Amount of Power: for Maximum Wire Size

From the maximum planned power transmitted we can calculate the maximum wire size needed. (Current = Power/Voltage; see volume 1.)

3. Acceptable Power Loss: for Minimum Wire Size

Larger wires have less power loss. We can calculate the minimum size wire needed for a stated acceptable power loss. (Recall: $P=I^2/R$.)

4. Mechanical Strength: for Minimum Wire Size

From the weight of wind, ice, and other objects on the wire, we can calculate the minimum wire size needed in order to support the total weight. (Books exist which provide all the necessary data and equations).

Understanding Wire Sizes

Introduction

If you are selecting or installing power lines you might need to read data regarding wire sizes and the resulting properties. Wires are measured in several different ways:

1. AWG (American Wire Gauge)
2. diameter, in inches
3. mils and circular mils

1. AWG (American Wire Gauge)

The AWG is a series of numbers, from 0000 to 30. Each number corresponds to a diameter. A higher number means a smaller diameter. (Stated the other way, a lower the number is a larger diameter). For example, a gauge #3 wire is thinner than a gauge #2 wire. A gauge #7 wire is thinner than a gauge #6 wire.

The reasoning for this method of sizing is due to the manufacturing process. In order to make the wire smaller we must extrude the metal one more time (an additional step), hence that wire is given a higher number in the American Wire Gauge sizing system.

When the "0" is used in the AWG sizing, more 0s designates a larger diameter wire. For example, gauge #000 is a larger diameter wire than gauge #00. Think of the 0s as going backwards in manufacturing time, to a larger diameter wire.

Note that when the initial numbering system was created the largest wire diameter was designated gauge #1. As larger diameters were added to the list, a creative numbering system was necessary. This is the reason for the gauge numbers which use "0."

Also note that the sizing with zeros may be listed in two forms, such as 0000 or 4/0. The "0" of the "4/0" means that the sizing is in the "0" range. The "4" tells us that there are four zeros. Hence, "4/0" is the same as "0000." Exact sizing data will be found in the Appendix.

2. Diameter, in Inches

This is the easiest form of wire measurement to understand. The diameter of each wire is measured in absolute terms, in the unit of inches. Wire diameters start from about .5 inches and go down to 1/100 of an inch. Exact sizing data will be found in the Appendix.

3. Mils and Circular Mils

A "mil" is not really a good scientific term, but it is very commonly used so we will work with it. Specifically: 1 mil = 1/1,000 of an inch. A related term is the "circular mil." The circular mil is a measurement of area. Specifically: 1 circular mil is the area made by a wire whose diameter is 1/1,000 of an inch. Exact sizing data will be found in the Appendix.

Choosing Stranding or Solid

Introduction and Stranding Process

"Stranding" means wrapping several wires together. In a stranded wire the finished power line has the appearance of twisted ropes. Stranding type conductors usually have 7 to 19 strands. All strands (each individual wire) carries the same phase of electrical current.

The process of stranding wire starts with one central wire. This wire is usually steel, which is used to provide the strength. Then other wires are wrapped around the central wire, all the way down the line. This gives the finished power line an appearance of twisted metal ropes.

Advantages of Stranding

With stranding we have the capability to create wires as large as we want. This is in contrast to individual electrical wires which have a practical maximum diameter: beyond a certain diameter an individual wire is not flexible, and it can break easily as a result of constant tension.

Therefore, several smaller wires are "stranded" together rather than using a single large wire. The net effect is a power line with sufficient strength of a larger wire, and current carrying ability of a larger wire, yet the greater flexibility of a thinner wire.

For example if we do a stranding of 7 wires each .5 inches in diameter then the overall diameter is 7 wires x .5 inches = 3.5 inches diameter total. The total strength of the power line is based on the diameter of 3.5 inches. The total current carrying ability is also based on the diameter of 3.5 inches. Yet the flexibility of the power line is based on the .5 inch wires, not the full 3.5 inches.

The disadvantage is expense. This process takes longer and therefore costs more than using a single wire of the same width.

Terminology

Note the terminology: Each individual wire is a "strand". The process of wrapping individual wires in a rope-like fashion is called "stranding". Any single power line which been stranded with individual wires is referred to either as a "stranded wire" or a "stranding".

Common Strandings

Most high voltage transmission towers hold three strandings (three power lines). Each stranding is wrapped with 11 or more strands. All strands in one power line carry one of the three phases.

As the power is delivered and is stepped down in voltage, the power lines have fewer strands. The individual strands may also be thinner.

Note that in underground cables these power lines are often "bundled" together. Bundling is the process of grabbing all three strandings and enclosing in a single plastic tube. (See chapter 9.8 for more details).

Chapter Summary

1. We can put the types of power lines into three broad categories:
   a. High Voltage Transmission Lines
   b. Community Level Distribution Lines
   c. Underground Cables

2. There are some issues common to all types of power lines.
   a. Right-of-Way
   b. Voltage, Current, and Power Loss
   c. Excessive Temperatures
   d. Choosing Which Wires to Use
   e. Above Ground or Below Ground

3. "Right-of-Way" refers to the area used by the utility company for their towers, poles, lines, and cables.

4. Tips when choosing right-of-way include:
   a. Place the lines along/below streets, highways, and railroad tracks.
   b. Choose the shortest paths possible.
   c. Coordinate with communication lines.
   d. Avoid crossing hills, ridges, and swamps
   e. Preserve the environment and scenic locations.

5. The right-of-way width for an electrical pole should be 15-50 feet. The right-of-way width for a tower should be 50-250 feet.

6. Temperature is an important factor in operating power lines. Important principles include:
   a. Wires expand and contract with temperature changes.
   b. Electric current creates heat. Therefore, wires must either be cooled or be designed to withstand high temperatures.
   c. Air naturally cools above ground wires. However, there is no natural cooling in underground cables.

7. Choosing which wires to use can be divided into these main categories:
   a. Choosing Which Voltage Wires to Use
   b. Choosing Which Material to Use
   c. Choosing Which Size of Wires to Use
   d. Choosing Stranding or Solid
   e. Choosing Insulation (how much, what types)

8. Choosing which voltage wire to use is not simple. Factors include: operating costs, power that might be lost, and the voltage of existing power lines.

9. Choosing the material of the wire involves factors such as conductivity, rust resistance, weight, strength, and insulation.

10. The most common materials for wires are copper, aluminum, steel, Copperweld, Alumoweld, and Aluminum–Conductor Steel Reinforced.

11. Choosing the size of the wire involves four factors:
    a. Line Voltage (minimum current, thus minimum wire size)
    b. Amount of Power (maximum current, and thus maximum wire size)
    c. Acceptable Power Loss (large diameter wires have less power loss)
    d. Mechanical Strength (minimum wire size to support weight)

12. Wires sizes are measured in several different ways: American Wire Gauge (AWG), diameter, mils, and circular mils.

13. The American Wire Gauge is a series of numbers. A higher AWG number corresponds to a thinner diameter.

14. 1 "mil" = 1/1,000 of an inch; 1 "circular mil" is the area made by a wire whose diameter is 1/1,000 of an inch.

15. "Stranding" is the process of wrapping several wires together. Stranding provides the strength of many wires, while keeping the flexibility of the individual wires.

16. Most power lines require some type of electrical insulation. This insulation is necessary to prevent power loss.

17. Air is a great insulator and is sufficient for most situations.

18. Power lines often need to be wrapped in insulating material, including rubber, thermoplastics, and taped insulation.

9.4

High Voltage Transmission Lines

Introduction

Transmission lines (also known as high voltage lines) are any wires which carry voltages of 70,000 Volts or greater. These power lines generally carry 100,000 to 1,000,000 Volts.

High voltage transmission lines carry the electricity from the power plant to the cities. The primary reason that we carry electricity at high voltages is to limit the amount of power loss.

As discussed earlier, delivering a given amount of power at a higher voltage uses less current. Using less current will result in less power loss. Therefore, using high voltages we can carry electricity over long distances, even hundreds of miles, with only minimal power loss.

The practical details of transmission lines which must be understood relate to two factors:

1. Potential for power loss along the line

2. Potential for damage to the line from the natural world

In this chapter we will look at some of the specific issues that relate to high voltage transmission lines and the towers which support those lines.

List of Topics for this Chapter
1. Types of Transmission Lines (categories of "high voltage")
2. Material of transmission wires
3. Electrical Insulation, Power Loss, Corona Discharge
4. Above Ground or Below Ground for Transmission Lines
5. Transmission Towers (High Voltage Towers)
6. Weather Tips for Transmission Lines and Towers
7. Safety of High Voltage Transmission Lines

Basic Types of Transmission Lines

Strictly speaking, the term "transmission line" is reserved for higher voltage wires that carry electricity over long distances. These lines are any wires that carry voltages of 70,000 Volts or greater. Transmission lines are also sometimes called "high voltage lines."

There are several varieties of transmission lines. These power lines are divided into categories, grouped by the range of voltage which the line carries:

1. HV = High Voltage: 70,000 Volts to 200,000 Volts
2. EHV = Extra High Voltage: 200,000 Volts to 800,000 Volts
3. UHV = Ultra High Voltage: 800,000 Volts to 1,500,000 Volts

Material of Transmission Wires

Wires for transmission lines are usually a combination of steel and aluminum. Aluminum is the lightest of the good electrical conductors. Therefore, aluminum is a good choice of material for hanging from the tall electrical towers. Aluminum also has good corrosion resistance. Once the aluminum oxide layer is created on the surface, then the surface does not flake or create holes.

However, these aluminum wires still need strength, and many of these "aluminum" wires are actually a combination of steel and aluminum. The innermost layers are steel, for strength, then outer layers are aluminum, which does the actual conducting of the electricity. (In some cases, both the steel and the aluminum conduct the electricity).

Electrical Insulation & Corona Discharge

In order to prevent electricity from conducting away from the wire (through other conductive materials nearby), we must have an insulator surrounding our wires. Air is a natural electrical insulator for power lines. Most of the time, air is not very conductive. Therefore, in above-ground power lines, air is often the only insulation that is needed. However, corona discharge can be a problem.

Corona discharge is a type of electrical insulation problem that contributes to significant power loss. This only happens with high voltage transmission lines.

In most circumstances, the air surrounding the wire is not very conductive, and therefore the electricity is not tempted to leave the wires. However, with very high voltages, particularly when carried through wires of small diameter, the electrical field from the wire tends to break up the air molecules. The air molecules are now charged. The air is no longer an insulator, but is more of a conductor. Some electricity will then conduct from the wire to the air. This process is known as corona discharge and can result in a significant amount of power loss.

To prevent this power loss from corona discharge, we can use larger diameter wires, use more insulation, and build taller towers (in proportion to the increased voltage).

Above Ground or Below Ground

Transmission lines are generally best placed above ground. Laying electrical transmission wires below ground is very expensive. The increase in cost is due to the additional technological issues, including: cooling requirements, high voltage dangers, and tunnel designs.

Cooling is the primary expense. When power lines are above ground, the air naturally cools the line. In contrast, when power lines are placed below ground the heat is not allowed to escape. Furthermore, with a higher voltage, a greater amount of heat will be produced. Without a cooling mechanism this heat will cause significant damage to the lines.

However, laying transmission cables below ground can be done. Several of the largest cities in the U.S. have been using underground transmission lines for decades. Furthermore, the technology for placing transmission lines below ground has gotten better in recent decades. For more details, see the chapter on Underground Cables.

Transmission Towers

Introduction

The electrical towers are tall metal structures which hold the high voltage transmission lines. The transmission towers carry large voltages from the energy source to the cities. The technical name for an electrical tower is "pylon"; however the term "tower" is equally acceptable. Towers for high voltage transmission lines can be built in many varieties. In this section we will look at some of the options.

General Structure of Transmission Towers (Fig. 9.7, Fig. 9.8, Fig. 9.9)

High voltage transmission lines can be put on either towers or poles, depending on the size of the power lines. Transmission lines (the wires which carry the highest voltages) are larger and heavier than all other power lines. Therefore the towers must be large enough to hold these heavier wires. Similarly, transmission towers must be sturdy in order to support these larger lines. Steel is the best material for the structure of the towers. Many towers also have concrete as the base.

An electrical tower has a minimum of three arms. Remember that electricity is generated in three phases. Each wire carries one of the phases, and thus each arm of the tower holds one of those wires, carrying one of those phases.

Sometimes you will see six arms rather than just three. Think of this as two 3-arm towers joined together. One side has three arms, which hold the three wires, and three phases. The other side also has three arms, which holds another set of three wires and its three phases. Therefore, over that same space of land, we have six transmission lines rather than just three.

Additional Wires on Transmission Towers

In addition to the 3 wires (or 6 wires) which carry the electrical power, there may be other wires on a transmission tower. A transmission tower may have a wire along the top (Figures 9.7, 9.8, 9.9). This wire is called the "earth wire", because it is connected electrically to the earth. This wire serves no purpose in transmitting the electricity. It does, however, have a very important purpose: lightning prevention. Lightning will hit the top wire, rather than the important wires carrying our power, and thus our power lines are not destroyed by lightning.

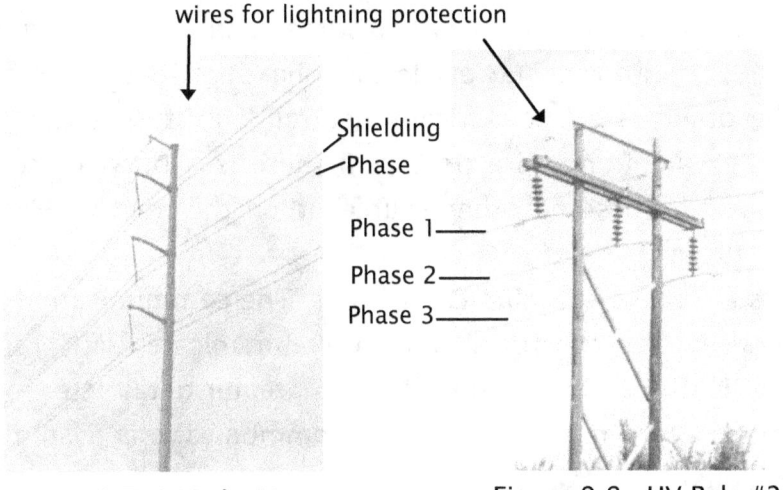

wires for lightning protection

Shielding
Phase

Phase 1——
Phase 2——
Phase 3——

Figure 9.7 HV Pole #1
115,000–160,000 Volts

Figure 9.8   HV Pole #2
115,000–160,000 Volts

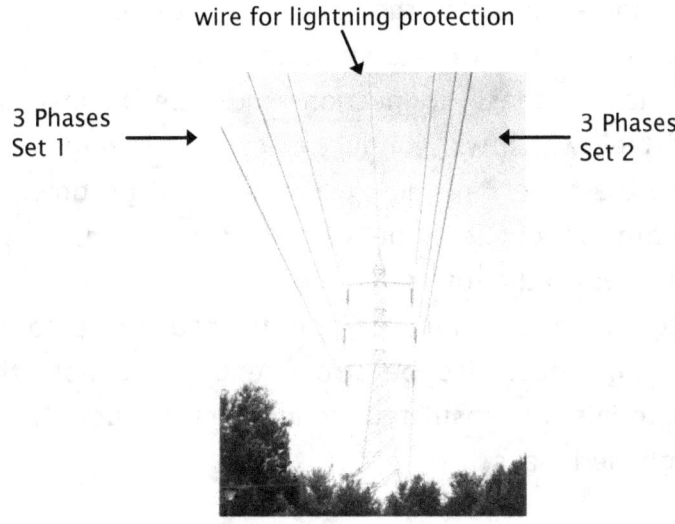

wire for lightning protection

3 Phases
Set 1

3 Phases
Set 2

Figure 9.9   EHV Tower
345,000 Volts and above

The tower might also have a shielding wire, running next to the power line (Figure 9.7). Often, each arm has two wires running close together. One wire carries the power (a particular phase of the transmitted electricity). The other wire is electrically neutral. This serves the same basic purpose as a lightning wire on top. If there is a burst of current (from lightning, or from some other source), then the neutral wire will take that extra current.

Many towers also have communication lines. The communication lines allow engineers in the control room to communicate with remote locations in the field. This will be discussed in greater detail later in this series. (See the chapter on Monitoring and Communications Systems, in the volume on Grids).

Connecting Transmission Lines to the Tower (Figures 9.7, 9.8, 9.9)

Power lines must be connected to the towers. This is not as simple as just tacking the power line to the tower, as there are particular requirements regarding insulation and strength.

First, the material for these connections must be an insulator. The tower is made of steel, which will conduct electricity. If we just attached the wire directly to the tower, all the electricity would go down through the tower. Therefore we connect the wires to an insulating material, which is then connected to the tower.

These insulated connectors must be able to insulate up to a million volts. The connections must also be strong enough to hold the large wires. Typical materials for insulated connectors include high grade porcelain and toughened glass.

Corona Discharge and Type of Transmission Tower

One method of limiting corona discharge is to use larger diameter wires. Larger diameter wires will necessarily require larger towers. Therefore, building taller towers gives us the ability to limit corona discharge.

Weather Tips for Transmission Lines and Towers

Introduction

Although we have mentioned weather and power lines elsewhere, in this section we present the main points regarding weather and transmission lines.

1. Build a Sturdy Tower

A sturdy tower is necessary to protect against heavy winds, including tornadoes and snowstorms. The tower can be made secure with a steel frame and a concrete base.

2. Install a Power Line which will Bear the Weight of Ice and Winds

Ice and wind add weight to the power line. Given enough ice or wind, over time, the power line will break. Therefore the wire must be strong enough to support all the possible ice and heavy winds in the area.

To know how much weight that the power line needs to support, and hence to know what size wire required, we use loading factors. These loading factors compensate for ice on wires and the weight produced by heavy winds.

There are several directories and websites available with loading factor regions drawn on maps. Simply find your area, get the loading factor, then add that weight to any other weight requirements of your power line. This will provide you with the total weight requirements for your power line, including factors of ice and winds.

3. Install a Top Wire to Prevent Lightning Damage

Transmission towers are usually the largest item in their area. They are also metal towers, with metal wires. This makes a natural place for lightning to strike. To prevent damage from lightning, a top wire is lined above all others. Lightning will hit this wire, the burst will be directed to the ground, and thereby prevent any damage to the towers or to the transmission lines.

4. Take Note of Winds, Eddy Currents, and Vibrations

All winds, even light winds, will form eddy currents around the wires. This can cause vibrations in the wire. Over time, the wire will deteriorate due to vibrations. Note also that steady light winds usually cause more vibration damage than sudden heavy winds. In order to prevent vibrations, we use vibration dampers. These dampers are placed around the wires and act as a cushion to soften the wind vibrations.

5. Adjust the Power Transmitted in Proportion the Outside Temperature

The temperature of the air around the transmission lines will limit the power that the wire can carry. If the wires are cooled by the air, then more power can be transmitted. If the temperature around the wires is warmer, then less power can be transmitted. If the total temperature of the wire becomes too hot then the wire may break. (See the chapters on Quality Control later in this series for more details). Therefore, we must never send more power through the line than the outside temperature will allow.

6. Clear Trees Nearby

The area around the towers must be cleared of trees. This is done in order to prevent trees falling on the tower, and prevent trees from being blown by storms into the tower. This will also prevent forest fires from reaching the tower.

Safety of Transmission (High Voltage) Lines

Introduction

There are two safety issues to discuss regarding transmission lines: the high voltage flowing through the wires, and the electromagnetic fields emitted from the wires.

Do not confuse the electric field with the voltage. The electric field is the field of electrical energy surrounding the wires. The voltage is the energy of the current flowing through the wires.

Hazards of High Voltage Electric Current

The voltage in a transmission line is very high, and therefore it is very dangerous for any person to touch these wires. Technicians who work on these power lines must take precautions. Specifically, the workers must connect their platform with the power line in such a way that the entire system is electrically neutral. As long as the worker is not connected to the ground then he is safe.

Clothing is another key to safety. The lineman wears non-conductive gloves, shoes, helmets, and other clothing so the current will not pass through his body. Linemen also use various poles and expandable tools to reach desired objects without touching the wires directly.

The wires are not usually a problem for the average person. The power lines are so high above ground that the average person cannot reach them. However, the power lines must be attached securely to the towers so that the lines will withstand heavy winds (and never be knocked down where people can touch them).

High voltage power lines should never be installed above schools or stadiums because a fallen wire at those locations could harm many people. These locations also provide easy access for adventurous teenagers to climb the towers, which could result in severe shocks or death.

Safety of Electromagnetic Fields (EMF)

There is often talk of effects of the electromagnetic fields (EMF) from high voltage transmission lines on human health. There is no data which proves that electromagnetic fields are harmful.

Numerous studies have been done over the past 25 years evaluating EMF and biological effects. One of the most comprehensive evaluations was the study mandated by Congress, managed by the National Institute of Environmental Health Sciences, over a period of several years (1992–1998). The net result from all these studies is that there is no correlation between EMF and cancers.

Several studies have also tried to create the results in the lab. For example, scientists would expose an animal or a human volunteer to EMF then observe any biological effects. The reports state that the supposed damage to health by power lines could not be reproduced by EMF exposure in the lab.

Furthermore, there have been studies done on people who work close to the lines and these studies show that the health of the individual is good. Specifically, studies have been done on linemen (people who work on power lines as a career), and farmers (who often have large towers in their fields). These studies show that you can work near these lines or with these lines for many years without any effects to your health.

However, it should be noted that the electromagnetic fields can induce electricity in another conductor (including the metal of a tractor). If that happens, a person working nearby can sometimes get a small shock, even when not touching the line. However, those who work on or near these lines say that this small shock is more annoying than harmful.

Furthermore, in addition to the official health studies of EMF, a brief look at the strengths of electromagnetic fields will tell us that the EMFs from power lines are relatively safe. Scientists have measured the EMF at various distances from power lines of all voltages. This data shows that the energy of the EMF decreases dramatically within a short distance. This extensive data can be summarized by two key facts:

•Within 100 feet of a power line, for the majority of power lines, the energy of the EMF decreases by approximately 85%.

•At 500 feet from the power line the EMF becomes negligible, even for the highest voltage power lines.

In conclusion, many research projects and extensive data tables exist, all of which demonstrate that EMF from high voltage power lines do not harm humans.

Chapter Summary

1. Transmission lines are high voltage power lines which carry electricity at 70,000 Volts and above.

2. There are three categories of high voltage transmission lines:
   a. HV = High Voltage: 70,000 Volts to 200,000 Volts
   b. EHV = Extra High Voltage: 200,000 Volts to 800,000 Volts
   c. UHV = Ultra High Voltage: 800,000 Volts to 1,500,000 Volts

3. The materials of transmission lines are usually aluminum and/or steel.

4. There are usually three transmission lines, one power line to carry each phase. In addition, there are usually other lines to prevent lightning damage and for communication.

5. Air is a natural electrical insulator and usually does a good job of preventing electricity conducting off the power line.

6. Corona discharge occurs when electrical current flows from the power line through ionized air. In order to prevent power loss from corona discharge, we can:
   a. Use larger diameter wires
   b. Put more insulation around the wires
   c. Build taller towers

7. Air naturally cools above-ground wires. This natural cooling allows more power to be sent than if these lines were underground. It is for this reason that high voltage transmission lines are usually above ground.

8. With a higher voltage, the towers must be taller. This is mainly because of the larger wires that are used. (Larger wires are used for both increased power and to reduce corona discharge). Furthermore, the height of the tower is directly related to the amount of voltage traveling through power line.

9. Towers must be built to withstand many weather conditions.

10. In order to withstand high winds, the towers must be built of steel and set into concrete.

11. Lightning is a significant concern because the towers are made of metal, the towers are usually the tallest structure around, and lighting has very high voltages.

12. The most common method of lightning protection for transmission lines is the earth wire. This is a wire lined along the top of the towers which conducts lightning directly to earth.

13. Wind adds a certain weight on the wires, and therefore the wires chosen must be able to support the additional weight.

14. Continuous wind also creates vibrations which can damage the wires over time. Vibration dampers can be used to protect the wires, acting as a cushion.

15. High voltage wires can be very dangerous. If you touch a live high voltage wire, you could be killed instantly.

16. The electric fields surrounding the power lines are not harmful.

9.5
Community Level Distribution Lines

Introduction

Distribution lines are the power lines we see in the neighborhoods lining the streets. These power lines start from the first voltage reduction, usually at a substation just outside the city. These power lines then proceed all the way to each home and business.

Note that the term "distribution line" is contrasted with the term "transmission line." Distribution lines carry 70,000 volts or less, whereas transmission lines carry 70,000 volts or more.

Most of the practical issues regarding community level lines are similar to those for transmission lines. However, there are a few practical points which differ.

List of Topics for this Chapter
1. Types of Distribution Power Lines
2. Distribution Lines: Practical Details
3. Poles for Distribution Lines
4. Trees and Tree Trimming
5. Inspections and Maintenance
6. Underground Cables

Types of Distribution (Community) Power Lines

There are three types of distribution lines, each connected in sequence. These types of lines and their terms are based on the stage of voltage reduction in the sequence, as well as what the wire connects. The three types of distribution lines are:

1. Primary wires, or simply "primaries"

2. Secondary wires, or simply "secondaries"

3. Service wires, or simply "services"

Primary wires are the lines that exist after the first voltage reduction [Fig 9.1(5)]. An example voltage is 69,000 Volts. Primaries connect the transmission substation which lies just outside the city to the distribution substation in the local neighborhoods. Primary wires are usually lined along highways or other major roads. Primaries can be lined above ground or below ground.

Secondary wires are the lines that carry the electrical current after the second voltage reduction [Fig 9.1(7)]. An example voltage is 13,800 Volts. Secondaries connect the distribution stations (the substations in the local neighborhoods) with distribution transformers (the box on a pole near the homes). Secondary wires are the wires that we see on wooden poles which line the streets by our homes. Secondaries can be lined above ground or below ground.

Services are the lines that exist after the third and final voltage reduction. Example voltage: 120 Volts. Services connect the distribution transformers (box on a pole near a few homes) with each specific home. [See Fig 9.1(9)], 9.5, 9.13].

Distribution Lines: Practical Details

The practical details of distribution lines are far simpler than for transmission lines. Distribution lines are usually made of copper, because copper is the most conductive of the practical metals. Distribution lines are also made of hybrid materials such as Copperweld.

Distribution lines are usually attached to poles along city streets or buried in conduits below the ground. At this stage we still have three phases of electricity, and therefore we have three wires (one wire for each phase) wherever these lines are placed.

Air acts as natural insulation for above ground distribution lines. We usually do not need any more insulation than the air.

The principles still exist for power loss. However, the circumstances differ. When power lines cross urban areas, we must consider the safety of the people as well as the power loss. Therefore, we reduce the voltage in stages. We want to keep the voltage as high as we can, and thus have minimal power loss, until safety concerns dictate that we reduce the voltage again.

Poles for Distribution Lines

Distribution lines are generally attached to poles. There are three arms, each holding one phase of electricity. Another wire is usually run along the top which serves to protect the pole and the power lines from lightning.

Poles are usually made of wood or steel. Wood poles are treated with a preservative, and are inspected regularly. Wood poles last about 10-20 years. Metal poles last much longer. However, they attract lighting much more easily.

The poles for distribution lines are usually lined along streets. This is most practical, for several reasons:

    a. This has less negative visual impact

    b. No new areas of land need to be set aside

    c. Easier access for maintenance and repairs

    d. Street lighting is more accessible to lines

Trees and Tree Trimming

Utilities do understand the value of trees. However, trees must be trimmed continuously or branches may fall on the wires during a storm.

Although the poles need far less of a right-of-way than transmission towers, the need for space is just as important. Trees often fall on the lines, particularly during a strong storm. For primary lines, we need to clear trees away from the poles, between 15 to 50 feet. For secondary lines, it is best to clear trees at least 15 feet away from the line.

Tree trimming is done very carefully. Electrical technicians go out and determine which trees need to be trimmed, then technicians mark each tree to be cut. Only when the trees are marked are the tree trimmers called into the area. Also, the tree trimmer is usually required to have written permission from the electrical technician before cutting any trees.

Inspections and Maintenance of Distribution Lines

Field inspections must be a continuous task because there are so many wires, poles, and transformers to inspect. The good news is that technology has become quite advanced. Modern technology has helped to make inspections easier and less time consuming. We will note a few of these devices in our discussion.

Wires can be inspected by "Thermovision equipment." This device uses infrared to locate hot spots on conductors. Any hot spot is an indication that too much power is being lost to the air. The wires can then be replaced or repaired as needed. Thermovision can be done from a van, a helicopter, or from the street.

Regarding poles, the inspection schedules are determined primarily by experience. The most important factors for inspection schedules of poles are the type of wood and the environment of the area. The strength of a wood pole can be determined by hitting the pole with a hammer. A decayed pole will sound hollow. Holes can then be filled in, and the outside can be treated with preservatives. Pole strength can also be determined by using a "Pole Test" device. In this device, a probe is put into the pole, and this probe is linked to a handheld computer. The operator keys in the type of wood and the diameter of the pole, then taps the pole. The probe then registers the resulting sound wave. This device will then automatically display the exact strength of the pole.

Underground Cables

Any of these distribution lines can be placed underground. Distribution lines are much simpler and cheaper to place underground than transmission lines. However, installing cables underground, even distribution lines, are still more expensive than installing overhead lines. There are also more technical issues to consider. The details of installing distribution lines underground will be discussed in the chapter devoted to underground cables.

Summary

1. Distribution lines are any power lines that carry 70,000 volts or less. These are the power lines which line the streets and travel through the local neighborhoods.

2. There are 3 types of distribution lines, each connected in sequence:
   a. Primary wires, or "primaries"
   b. Secondary wires, or "secondaries"
   c. Service wires, or "services"

3. Primaries connect the transmission substation, which lies just outside the city, to the distribution substations in the local neighborhoods. Primary wires are usually lined along highways or other major roads.

4. Secondaries are the lines that carry electrical current after the second voltage reduction. Secondaries connect substations in the local neighborhoods to the box on a pole near the homes. Secondary wires are the wires that we see on wooden poles which line the streets by our homes.

5. Services are the lines that exist after the third and final voltage reduction. Services connect the distribution transformers (the box on a pole near a few homes) to each specific home.

6. Any of these distribution lines can be placed underground. Installing distribution lines underground is much simpler than for transmission lines. However, installing cables underground is more expensive than installing overhead lines and there are more technical issues to consider.

7. Distribution lines are usually made of copper. Distribution lines are also made of hybrid materials such as Copperweld.

8. Air acts as natural insulation. For distribution lines, we usually do not need any more insulation than the air.

9. Poles for distribution lines are usually made of wood or steel. Wood poles are treated with a preservative. Both wood and steel poles are inspected regularly.

10. The poles for distribution lines are usually lined along streets. This is most practical, for several reasons:
    a. Less negative visual impact
    b. No new areas of land need to be set aside
    c. Easier access for maintenance and repairs
    d. Street lighting is more accessible to lines

11. Trees must be trimmed regularly or branches may fall on the wires during a storm.

12. Wires can be inspected by Thermovision equipment. This device uses infrared to locate "hot spots" on conductors. Thermovision can be done from a van, a helicopter, or from the street.

13. Wood poles should be treated with a preservative when they are put up. Inspection schedules are determined primarily by experience. The most important factors for inspection schedules of poles include the type of wood and the environment of the area.

14. The strength of a wood pole can be determined by hitting the pole with a hammer. A decayed pole will sound hollow. Pole strength can also be determined by using a Pole Test device. This device can automatically register the exact strength of the pole based on acoustics. Holes can then be filled in, and the outside can be treated with preservatives.

9.6

High Voltage Direct Current (HVDC) Power Lines

Introduction

Electric power can now be carried over High Voltage Direct Current transmission lines (HVDC). The HVDC lines are best used where there is fixed distance of great length over which we wish to transmit power. The technology is relatively new. At this time, approximately 100 HVDC transmission lines exist world-wide.

The HVDC technology is growing in popularity around the world. However, the DC transmission lines are often sold improperly. Governments can be mislead into buying HVDC transmission lines rather than AC transmission lines for the wrong reasons. This results in more power loss, not less. This can also result in paying more money than necessary. In this section we will provide an overview of HVDC power lines, followed by a discussion of where HVDC power lines are most effective.

List of Topics for this Chapter
1. Basic DC Transmission System
2. HVDC Advantages and Limitations
3. HVDC Lines used Under Water
4. HVDC over International Borders

Basic DC Transmission System

1. A conventional AC generator is used to create alternating current, in three phases.

2. A conventional step-up transformer is used, to raise the AC voltage to high amounts (such as 500,000 Volts).

3. The high voltage AC is converted to high voltage DC.

   A typical mechanism involves only using the forward push of the AC, but blocking the effect of the reverse pull. This will create current in only one direction within the DC wire. Another mechanism involves taking the reverse pull of the AC and routing it in such a way as to become a separate DC current.

4. The high voltage DC is carried over DC transmission lines.

   The mechanism for current is still electrons bumping down the line, with the initial push coming from the cyclic alternating current. The only difference is that the reverse pull is never applied to the DC wires.

5. At the other end of the line, the high voltage DC is converted to high voltage AC. This is usually done in an inverter, similar to those used with solar cells.

6. The high voltage AC is sent to conventional substations, where it is lowered in voltage, and sent along distribution lines (same as the normal distribution process for AC).

HVDC Advantages and Limitations

The most common selling points presented by the HVDC industry include:

- High Voltage DC lines have less power loss than AC lines
- High Voltage DC lines are cheaper than AC lines
- High Voltage DC lines can be used for long distances
- High Voltage DC lines are better than AC lines for use under water
- High Voltage DC lines must be used between nations

While there is some truth to each of these claims, these statements can be misleading. The whole truth must be understood if we are to make intelligent decisions. The truths regarding HVDC are the following:

1. The Power Loss for HVDC and AC Lines are Essentially the Same

The primary amount of power loss is related to the inherent resistance in the material of the wire. The metal used for the power line is the same in AC and DC transmission lines. For example, aluminum might be used in either wire. Therefore, the power loss per mile of wire will be identical in either the AC or the DC transmission line.

2. Power Factor May be Better in DC lines

The AC transmission line does produce a small amount of power loss which does not occur in DC transmission lines.

This type of power loss is related to the "Power Factor". The details of the Power Factor will be discussed later in this series, but for the moment we can say the following: Power Factor is essentially a type of efficiency. This efficiency is associated only with the transmission of electrical power. There are several causes for a low power factor (the most common is phase difference between current and voltage) and several methods to improve the power factor. Note also that most of these problems are commonly fixed at substations.

DC lines are less prone to this type of efficiency loss than AC lines. Therefore, if none of the methods to improve the Power Factor in AC power lines are an option, then using the HVDC lines may be a sensible decision. Similarly, if the methods are not producing the desired results, or if the methods cannot achieve further efficiency, then using the HVDC lines may provide greater efficiency than traditional AC lines.

3. It is Always Best to Build Short Transmission Lines

With a shorter line, whether AC or DC, less power will be lost (due to the inherent resistance in the wires). Therefore more power will reach the final destination with shorter lines, regardless of AC or DC.

Similarly, a longer line is rarely more efficient than a shorter line, regardless of the AC versus DC. Therefore, due to the inherent resistance in the wires, the power loss in a long HVDC line will likely be greater than the power loss in a shorter AC line.

4. HVDC Lines are Best Used Where Fixed Distance of Great Length

Power loss due to the causes of the Power Factor will compound with distance. For example, the phase differences between voltage and current will grow as the power travels down the line. This results in larger amounts of power loss at each mile. Furthermore, without a substation to repair a power factor problem, any cause of power factor will gradually escalate.

Therefore, in locations where it is physically impossible to shorten the distance of the transmission line or to add a substation then a HVDC transmission line may be a sensible option. Such locations include large bodies of water or complex terrain.

5. Cost–Benefit: Efficiency versus Cost

HVDC is generally more expensive to build than an AC system. Therefore, only when the cost savings from increased efficiency outweighs the additional cost of installation will the HVDC line be economically sensible.

6. HVDC Lines are Best Suited in the Following Situations

HVDC lines are best suited where there is a fixed distance of great length, or where quality control is difficult to maintain. Therefore the following are the best uses for HVDC power lines:
   a. Transmitting power across bodies of water
   b. Transmitting power across national boundaries

Each of these best uses of HVDC will be explained in greater detail below.

HVDC Lines Under Water

One of the truly most effective uses of HVDC lines is for transmitting power across bodies of water. Remember the main HVDC principle: HVDC lines are best used where there is fixed distance, of great length, over which we wish to transmit power. In the case of water, we cannot shorten the distance. Nor can we put a power plant or substation in the middle of the water. If we cannot shorten the distance, then DC transmission lines may be an option.

Using HVDC rather than AC would save on that additional power loss. Furthermore, if the body of water is large then the distance is long enough for DC transmission lines to be economical. Therefore, large bodies of water are the most sensible location for High Voltage DC Transmission lines to be used. In the United States, such locations include: the Great Lakes, the San Francisco Bay, and the Hawaiian Islands.

HVDC Across International Borders

The other effective use of HVDC is when power must be transmitted across international borders. The reasons are related to the quality of the electricity.

The standards and quality control for electrical power can differ from country to country. When these differing national systems interconnect, the quality of the power can suffer tremendously.

This is particularly true of frequency. When electrical systems interconnect the frequencies must be the same. However, the specific frequencies used by each country may differ. In addition, the frequency can change or create harmonics along the line. Consequently, if the frequencies from different systems are not aligned, then the efficiency will drop significantly, and sometimes cause equipment to fail.

Using HVDC eliminates this problem. Direct current has no frequency. Without frequency, there can be no quality issues regarding frequency. Therefore, many nations choose to transmit power across national borders using HVDC rather than AC. After the power is in the second country, that second nation will convert DC to AC using its own standards.

Summary

1. High Voltage Direct Current (HVDC) is an alternate method of transmitting high voltage power. The technology is growing in popularity, but is often chosen improperly.

2. The basic HVDC system is as follows:
   a. Alternating current is created in a conventional generator.
   b. The voltage of the AC is increased using a conventional AC transformer.
   c. High voltage AC is converted to high voltage DC.
   d. The high voltage DC is carried over DC transmission lines.
   e. The high voltage DC is converted to high voltage AC.
   f. The high voltage AC is sent to conventional substations.

3. High voltage direct current transmission lines might truly be the best choice in the following situations:
   a. Transmitting power across national boundaries.
   b. Transmitting power across bodies of water.

4. Regarding power loss, high voltage DC lines are not any better than high voltage AC lines. Furthermore, if DC lines are used incorrectly then the power loss can be significant.

5. Using a longer the transmission line, whether AC or DC, more power will be lost. This is due to resistance inherent in the wire. Therefore, it is always best to build short transmission lines.

6. One of the most effective uses of HVDC lines is for transmitting power across large bodies of water. This is because:
   a. the distance is fixed (it cannot be shortened)
   b. the distance is long enough to make HVDC economical

7. HVDC is also effective when transmitting power across international borders. The reasons are related to the quality of the electricity in each country, particularly regarding the frequency.

9.7
Underground Electrical Cables

Introduction

Underground electrical systems look nicer and give a cleaner appearance. However, they are more expensive and they have their own technical issues. Underground cables are best used for the carrying electricity within the cities as distribution lines rather than as the high voltage transmission lines. In this chapter we will discuss the basic concepts and issues related to underground electrical cables. In the next chapter we will discuss practical details on cable design and laying cable.

List of Topics for this Chapter
1. Overview of Underground Cable Systems
2. Where Cables are Suitable
3. Geology and Underground Cables
4. High Voltage Cables as Underground Power Lines
5. Underground Cable vs. Above Ground Wires (Pros and Cons)

Overview of Underground Cable Systems

Basic Types of Underground Cable Systems

There are two basic underground cable systems: 1) the Direct Bury System, and 2) the Submersible or Conduit System.

In the direct bury system we dig a trench, then bury the cable. The final transformer is usually a box above ground. The submersible/conduit system is a system of pipes underground. The final transformer is usually placed in a room underground.

The conduit, being a pipe, has much greater physical support and greater resistance to water than the direct bury method. However, the conduit system is much more expensive to install.

Pad Mounted Transformer (box above ground)

The final transformer is usually an item called a "Pad Mounted" Transformer" (Figure 9.10). The Pad Mounted Transformer is a large metal box which sits above ground. The "pad" refers to the concrete pad that the transformer box sits on. If you look around at many businesses you will see these metal boxes. These are the transformers of the final stage of the sequence, very much like the transformers on poles for above ground systems.

Leading to Above Ground

At some point, the electrical wires must come from below ground to above ground. There are essentially two options: 1) use a special device to connect the cable with the above-ground wire, or 2) keep using the cable, just simply hang it above ground. The first method uses a connecting device called a "riser." The riser connects underground cables to overhead lines. The second method is to simply hang the cables above ground. When the cable (previously underground) is hanging above ground like traditional wires, this is known as an "aerial cable."

Fig 9.10 Pad Mounted Transformer

Where Cables are Suitable

Cables are most suitable where there is significant population density, or where the hazards are greater above ground than below ground.

Because the cost of underground cables is high, there must be a sufficient population density. In practical terms, sufficient population density means that there are enough users in a small area to make laying the cables cost effective. This means that cables are best suited for apartment complexes, hospitals, shopping centers, large businesses, and manufacturing facilities.

Another factor when choosing whether to place power lines above ground or below ground is to consider the relative hazards from nature. If there are significant hazards above ground, then putting power lines below ground is a good idea. Regions which are subject to frequent tornadoes, hurricanes, or blizzards are best served by power lines below ground. This criteria applies equally to transmission lines as well as distribution lines. Regarding transmission lines, note that the potential damage to from frequent tornadoes or blizzards is often a greater factor than the instillation cost.

However, if the hazards are greater *below* ground than above ground, then it is best to leave power lines above ground. Regions which are subject to earthquakes, landslides, or flooding are best served by keeping power lines above ground. Similarly, areas with swamps, sinkholes, or unstable soils are best served with power lines above ground.

Geology and Underground Cables

Introduction

If we are considering the option of placing power lines underground, then we must examine the geology in that area. We must know exactly what we are putting the power lines into so that we can protect the lines from potential damage. Furthermore, some regions are not suitable for any underground cables. In this section we will examine some geological characteristics to consider before installing power lines underground.

Soil Conducts Electricity

The earth conducts electricity very well. To get an idea of how well the earth conducts electricity, consider how we use this property of the soil for practical purposes. For example, devices for lightning protection (such as lightning rods) are connected to the ground. Electricity that has been used in our homes is sent into the ground. Some radio stations boost their power by installing their transmitting antennas within conductive soils in the ground.

However, this same electrical conductivity is something that we do *not* want when we have power lines underground. Therefore, it is essential that all power lines which are placed underground are wrapped in a high quality electrical insulator.

Temperature Below Ground

There is no air to circulate underground. Power lines create heat as electricity travels through, but there is no way to dissipate that heat. The power lines get hotter and hotter, until the heat is too much and damages the cables. Therefore it is important to provide some cooling mechanism for underground cables. At the very least, allowing a path of air to travel around the cables will help. The most effective method for cooling underground power lines is to send a liquid or gas coolant past the cable.

Water in the Soil

Water is a major concern. In fact, water damage is the number one cause of failure for power lines underground. The most significant problems caused by water are a) corrosion of the power line, b) conductivity of water, and c) shifting soils.

a. Corrosion: Water can cause metal to corrode. More specifically, the ions in the water will corrode the metal of power lines. This also means that water can corrode the outer layers of cables, particularly metal jackets.

b. Conductivity: The ions in the water conduct electricity. The net result is that the water can draw current away from the power line, resulting in significant power loss.

c. Shifting Soils: Water can also cause soils to be unstable, causing the soils to shift. The shifting soils can then damage the power lines.

There are two key methods that are used to protect underground cables from water damage:

1. The outer layers of the cable must be resistant to water.
2. Water must be led away through engineered channels.

Corrosive Soils

In addition to water being corrosive, the soil can also be corrosive. Acids in the soil can destroy metals, including power lines. The acidic soils can also corrode the metal jackets. Therefore, all underground cables must have an outer layer which will protect the power lines from corrosive soils.

Conductivity in the Soils

In addition to the ions in the water conducting electricity, the soil also contains many ions which will conduct electricity. (These ions are what makes the soil such an effective electrical conductor as discussed above). Therefore electrical current will often choose to deviate from the power line and travel through the soil, rather than continue down the power line. This results in significant power loss.

To prevent this, the cable must have a layer which is an electrical insulator. This electrical insulator will physically prevent any current from traveling beyond the cable.

Rocky Soils

Rocky soils can damage cables. Rocks are abrasive, therefore these rocks can puncture and tear the outer layers of the cable over a period of time. Earthquakes, landslides, and other geologic events can cause the earth to shift dramatically. When the earth moves, the cable buried beneath the earth also moves. If the change is drastic, then the cable is likely to break. Damage from rocky soils can be minimized by choosing locations carefully, and with the use of a conduit (see the next chapter).

Shifting Soils

Shifting soils cause another issue: unburying the cables. In some cases, the cable is intact but the ground on top is shifted away. Eventually the cables may become unburied. The chances for accidental damage from weather, cars, and construction equipment become much greater. This situation occurs to the greatest degree in sand or other fine soils.

Damage from shifting soils can be minimized by avoiding certain locations (such as fault lines and water tables), as well as by burying the cables to a sufficient depth.

High Voltage Underground Cables

It is possible to place high voltage power lines below ground. Several major metropolitan areas have placed high voltage lines underground with success. This is very expensive and many steps must be taken, but it can be done. The process is similar to the underground cables used in distribution lines. However, more insulation and more cooling are required. These cables must also be laid much deeper and have much greater mechanical strength.

The high voltage cables must be put into a pipe (like the conduit). For additional protection the pipes can be laid in a tunnel. Thus the entire cable system is the following: high voltage wires, wrapped in insulation, which is wrapped in a cable jacket, then set into a pipe, and placed inside a tunnel.

The high voltage cables must also be cooled. This is done by sending coolant through the pipes, which flows around and along the cables. The two most common coolants are: oil and SF_6. (The SF_6 is usually sent as compressed gas).

Underground Cables vs. Above Ground Wires

Introduction

Choosing to place power lines above ground or below ground is a complex decision. The final choice is usually partially scientific and partially subjective.

Above Ground Wires: Advantages

1. Cheaper
2. Natural electrical insulation
3. Air naturally cools
4. No damage from water
5. Can be put to residential and less populated areas
6. Easier access for inspection and maintenance
7. Distribution lines can be placed along roads, no new land needed

Above Ground Wires: Potential Problems and Disadvantages
1. Potential damage from wind, ice, snowstorms, lightning
2. Potential damage from trees, trees must be trimmed
3. Overhead lines are less attractive

Underground Cables: Advantages
1. No damage from high winds, ice, snowstorms, or lightning
2. No damage from trees
3. No need to cut trees down or trim trees
4. Sensible for cities and dense populations
5. Sensible for businesses, manufacturing, hospitals, apartments
6. Cleaner look
7. No new right of way needed, no new land cleared

Underground Cables: Potential Problems and Disadvantages
1. Water damaging the lines
2. Drainage design issues
3. Extra material needed to prevent water damage
4. Corrosion
5. Cable dig-ins
6. Innocent digging can damage cables
7. Insulation failures, and extra material to prevent such failures
8. Overheating causes serious damage, a cooling system is required
9. Requires more material: more insulation; more mechanical strength; more physical support
10. Requires greater separation from other underground structures
11. More difficult to inspect and repair
12. Costs more

Possible Reasons to Invest in Underground Cables
1. City ordinances
2. Cost of purchasing right-of-way above ground is too expensive
3. Appearance of the area
4. Congestion of the area, population density
5. Tornadoes, hurricanes, or blizzards are common to the area

Chapter Summary

1. Underground cables look nice but there are many technical issues to consider.

2. The temperature of underground power lines will increase to the point of damaging the line. Therefore an artificial coolant should be added.

3. Ions in the soil and water will conduct electricity away from the power line, therefore cables must have a layer of electrical insulation.

4. Water and soils will corrode the power lines. Therefore cables must have a layer of protection which will prevent water and acidic soils from damaging the power line.

5. Rocky soils will damage cables. Therefore the outer layer of cable must be resistant to abrasion, and may need to be placed in a conduit.

6. Shifting soils will leave cables vulnerable to damage from the outside, and therefore must be buried to a sufficient depth.

7. Placing high voltage lines underground requires tunnels, large conduits, and a constant flow of coolant.

8. Underground cables are generally more expensive than above ground lines, due to the number of extra layers required for protection, as well as for the cooling mechanism.

9. A geographic region should have a sufficient population density in order to justify the cost for underground cables rather than overhead lines.

10. Choosing to place power lines above ground or below ground is partially scientific and partially subjective. Most of the factors are related to the weather and geology in the region. Other factors include right-of-way logistics and cost.

9.8
Cable Design and Laying Cable

Introduction

In this chapter we will look at the practical details of cable design and installation. An underground cable needs much more than just the conductive wire. An underground power cable also needs insulation, protection, and mechanical strength. These requirements are achieved through effective design.

Similarly, the method of laying cable is very important because laying cable improperly can lead to early damage, while laying cable properly can protect the cable for much longer periods of time.

There are two basic underground cable systems: the Direct Bury System and the Conduit System. We will examine the practical details for laying cable in each of these two systems.

List of Topics for this Chapter
1. Basic Cable Design
2. Advanced Cable Design
3. Laying Cable Overview
4. Direct Bury Method
5. Conduit System Method

Basic Cable Design

An underground cable needs much more than just the conductive wire. A cable needs multiple layers of protection in order to prevent damage from anything in the ground. The basic parts of the cable include: 1) conductor, 2) insulator, 3) bundling, and 4) jacket.

The conductor is the actual wire that carries the electricity. Each conductor is wrapped in an electrical insulator. The layer of electrical insulation is necessary so that power is not lost to the surrounding area.

The bundling is three wires together. Remember that we have three wires, one wire for each phase of electricity. These three wires, each insulated, are bundled together.

The bundle of three wires is put into a jacket, which is the final protection from the outside elements. The jacket is the outermost layer. The jacket exists to protect the cable from water, debris, and other weathering effects. In some cases the jacket also provides strength.

Advanced Cable Design

Introduction

Due to all the protection that is required for an underground power line, cable designs are usually more advanced than the basic design described above. A typical advanced cable design will have these items, in the following order (Figure 9.11):

1. Conductor
2. Shield
3. Insulation
4. Semiconducting Material
5. Metallic Shielding
6. Bedding Tape
7. Cable Jacket

Figure 9.11 Cable Design

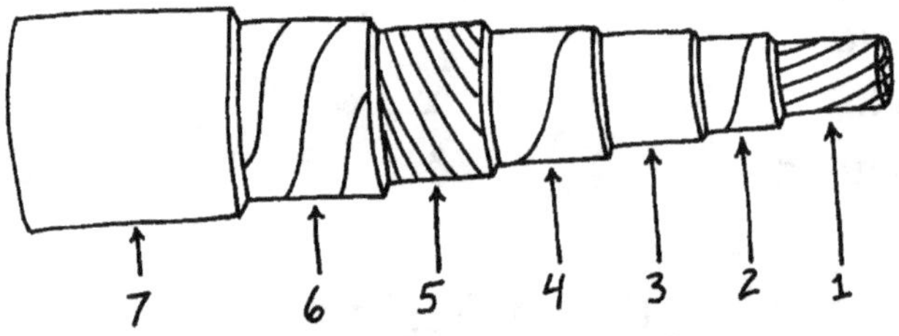

7) Cable Jacket 6) Bedding Tape   5) Metallic Shielding
4) Semiconducting Material 3) Insulation
2) Strand Shielding 1) Conductor

Of these layers, we will look most closely at the conductor, the insulation, and the jacket.

Conductors

The initial conductor is a wire which carries electrical current. The conductors of underground cables are usually made of copper or aluminum.

Several of these wires are usually stranded together (given the woven rope appearance as described earlier). The total width of each stranded wire will dictate how much current can travel through that line.

Each stranding is a single phase. Most cables have three strandings, one for each phase. Some cables have more strandings (similar to the six arm transmission tower described earlier).

Each stranding is then wrapped in electrical insulation (see the section below). In advanced cable design, the stranded conductors are first wrapped in a strand shielding, then in a layer of insulation.

Insulation

Electrical insulation is very important in underground cables. Recall that in power lines placed above ground, the air acts as a natural insulator. We do not have that natural insulation in underground cables, and therefore we must provide our own electrical insulation. Materials for electrical insulation can be put into the following categories:

1. Taped insulation
   a. Impregnated paper insulation
   b. Varnished-cambric insulation

2. Rubber insulation
   a. Oil based rubber
   b. Butyl synthetic rubber
   c. Silicone Rubber
   d. Ethylene-Propylene Rubber-base (EPR)

3. Thermoplastic insulation
   a. Polyvinyl Chloride (PVC)
   b. Cross-linked polyethylene

Note that the insulation wears over time. The insulation of all cables must eventually be replaced. Therefore, all cables must be dug up after a number of years in order to replace the insulation.

Bundling and Filling in Spaces

Conductors are usually a bundle of wires. As stated above, for each phase of electrical current we create a stranded wire (several wires twisted like rope). Then we insulate each stranded wire. At this point, we take all stranded wires, each layered in insulation, and clump them together. (Think of grabbing three computer cables in your hand). This procedure is known as "bundling".

Note that there are spaces between the insulated strands in the bundling. These spaces may seem small, however these spaces are large enough for water to sit or to travel. This water can then cause significant electrical failure anywhere along the line. Therefore these spaces should be filled. Alternately, encase all wires in a water resistant material (such as an additional shield or jacket).

Cable Jacket

The cable jacket is the outside covering of the cable. It is the first wall of resistance to everything potentially damaging from the environment (including water, rocks, and corrosive soil). The jacket also provides strength, flexibility, and resistance to abrasions.

Therefore, the material we choose is very important. Sometimes multiple jackets are used in order to provide multiple layers of protection. The materials for jackets can be put into the following categories:
1. Fibrous jacket
2. Rubber jacket
3. Thermoplastic jacket: PVC or Polyethylene
4. Metallic jacket: Lead or Aluminum

Note that the metallic jacket is often used as a second jacket. The metal jacket is often placed over the top of the fibrous, rubber, or thermoplastic jacket. The metallic jacket is used to provide extra mechanical strength. It is also used to prevent water from coming into the wire.

Overview of Laying Cable

As stated above, there are two basic methods to laying underground cable: the Direct Bury Method and the Conduit Method. In the direct bury method we dig a trench then bury the cable. The submersible or conduit method is a system of pipes underground.

Because the conduit is a pipe, it has much greater physical support and greater resistance to water than the direct bury method. However, the conduit system is much more expensive to install.

In either case, it is important that we protect the underground cables from being damaged. In the next sections we will discuss some practical tips regarding laying cable in both of those underground systems.

Direct Bury Method

Introduction

In the direct bury system, the cable is put directly into the ground. The cable is then carefully covered up. A related note on conduits: when burying cables directly is not advisable for a particular reason, you may be able to use conduits instead. Although conduits are not part of the direct laying cable system, we will mention their use as appropriate.

Practical Considerations when Laying Cables in Direct Bury Method

1. Cables must be buried to a depth of at least 2 feet.

2. Cables must be put deeper depending on the voltage:

Voltage	Minimum Depth
600 Volts or less	24 inches (2 feet)
601 to 22,000 Volts	30 inches (2 1/2 feet)
22,001 to 40,000 Volts	36 inches (3 feet)
40,001 Volts and above	42 inches (3 1/2 feet)

3. Where the area experiences frost, the cables must be placed several inches lower than the minimum depth for voltage listed in tables.

4. Cables must be laid as straight as possible.

5. However, cables should *not* be laid under major roads. The weight of the vehicles will damage the power lines over time.

6. Options for laying cables straight but not under roads include:
    a. Lay cables in back lots. These include the back yards of homes, and the less traveled back areas of hospitals and businesses.
    b. Lay cables in the ground adjacent to street. These cables are laid under the grass between sidewalks and businesses, rather than under the street.
    c. Use a conduit. If the path is necessary to put under major streets and highways, then a conduit (pipe) must be used.

7. Curves should be avoided. If the path for a cable must curve, then a large radius must be used in order to prevent the cable from breaking.

8. Do not lay cables under businesses or other buildings.

9. Avoid soil that is unstable or hazardous. These types of soils include mud, shifting soils, and corrosive soils. If the path must pass through these soils, then a conduit must be used.

10. Avoid rocky soils. Rocks can damage the outer layers of the cables.
    If you must put cables in rocky soil then you can smooth out the trenches and then add softer dirt to the trench before laying cable. You can also build a conduit.

11. Prevent water damage. This is very important – the greatest hazard to underground cables is water. Careful planning and design must be done to prevent water damage. Tips include:
    a. The cable itself should have great protection. This includes putting the wires in a jacket or sheath which houses all the cables.
    b. Avoid laying cables near sources of water.
    c. Put cables in conduits.
    d. Angle the path so that water drains off.
    e. Design paths of cables and conduits so that drainage of water leads to sewer lines.

12. Cables must be placed 5 feet from swimming pools.

13. In earthquake areas, cables must be very flexible.

14. Try to avoid putting cables above or below any other underground systems and structures (such as gas lines, subways, and foundations for buildings).

   If cables must be placed near other underground items then three items are required:

   a. Minimum of 12 inches between each underground system.

   b. Each underground item must be well supported.

   c. There must be access to each item, without disturbing another.

15. Electrical cables *can* be placed with communication cables.

   This is the notable exception to the rule above. Electrical cables and communication cables can be placed next to each other. They can be placed on the same level, and no separation is needed. However, agreements needed to be made with all appropriate companies.

16. Fill the trenches very carefully. Cables can be easily damaged by rocks. First put sand on the cables, then put a final layer of dirt on top the sand.

Conduit System Method

Introduction

   The conduit is essentially a tunnel or a pipe. The cable is laid inside the tunnel, and this tunnel with the cable is all underground.

   The conduit has much greater physical support than the cable alone. Therefore this conduit can be used where mechanical strength is an issue. Also, since the conduit is a pipe it has greater resistance to water than the cable alone.

   It is always best to lay cables directly when possible, because that method is much cheaper than the conduit method. However, sometimes the direct bury method does not provide enough strength or protection, and the region cannot be avoided. In these cases, the conduit method becomes a sensible solution.

In addition to the conduit itself, the conduit system uses two other items: vaults and manholes. Vaults are rooms underground. The vault is where the transformers are located. The manhole is where the electrical workers can access the conduits and vaults.

There are many practical points when laying underground electrical cables using the conduit method. Note that most of the practical points listed above for the direct bury method apply just as much to the conduit method. Those points will not be repeated. Additional points that relate specifically to the conduit method are listed below.

Practical Considerations when Laying Cables for Conduit System
1. Conduits can be made of many materials, including concrete, steel, plastic, and fiberglass.

2. Conduits should be no less than 2 feet from the surface.

3. Conduits should be short and straight, 500 ft length maximum. The end of each conduit is a manhole, where the cables are spliced together.

4. Conduits can be placed under streets. This contrasts with the Direct Burial Method, where cables should not be under the streets.

5. Conduits should be sloped so that water drains. Often the conduits are sloped both ways, with a gradual curve. This allows water to drain toward the manholes at either side.

6. Manholes can be placed in the middle of intersections or along sidewalks.

7. Manhole drainage: Often the bottom of the manhole is dirt, with rocks on top. This arrangement allows water to drain to the earth.

8. Many factors of direct burial are also applicable to conduits. Read the earlier sections on Direct Bury Methods for details.

Summary

1. The basic parts of a cable are the conductor, the insulator, the bundling, and the jacket.

2. Advanced cable design will have the following items: conductor, shield, insulation, bundling, semiconducting material, metallic shielding, bedding tape, and cable jacket.

3. Conductors in underground cables are usually stranded. The conductors of underground cables are usually made of copper or aluminum.

4. Stranded conductors are wrapped in insulation. Each stranded conductor carries one phase of electricity. These insulated conductors are bundled together, and often encased in a shield or jacket.

5. The cable jacket is the outside covering of the cable. It is the first wall of resistance to everything potentially damaging from the environment. The jacket also provides strength, resistance to abrasions, and flexibility.

6. The metallic jacket is used to provide extra mechanical strength and prevents water from coming into the wire. However, metallic jackets can corrode in acidic soils.

7. There are two general methods for installing power lines underground: the direct bury method and the conduit method.

8. In the direct bury method we dig a trench then bury the cable. The direct bury method is simpler and cheaper, but this method is not as durable.

9. The conduit method is a system of tunnels underground. The cable is laid inside these tunnels. The conduit system requires a more advanced design and is more expensive to install, but the power lines will last longer.

10. There are many important practical points regarding laying underground electrical cables. Some of the more important points are:

    a. Cables and conduits must be buried to an appropriate depth
    b. Cables and conduits should be as straight as possible
    c. Cables and conduits should not be placed near water
    d. Cables should not be placed under streets or buildings
    e. Conduits should have an adequate system of drainage

9.9
To the Homes and Businesses

Introduction
In this chapter we will look at the final stages of electricity, particularly the delivery of electricity to homes and businesses. We will also discuss the use of electricity in an appliance, and what happens after electricity has been used for the desired purpose.

List of Topics for this Chapter
1. Sequence of Electricity from Substation to Your Home
2. The Final Transformer and the Three Service Wires
3. Sequence Through an Appliance
4. Neutral Wires and Ground Wires
5. 120 Volts, 240 Volts, Prongs, and Outlets
6. Colors of Wires
7. Fuses and Circuit Breakers
8. Two Phase and Three Phase Wiring
9. 3-Phase Wiring vs. 3 Wires of a Single Phase
10. To the Businesses

Sequence of Electricity from Substation to Your Home

The sequence of electricity from the local substation to the home is relatively straightforward. Note that the sequence for an underground cable system is essentially identical to the above ground system; the only difference is we don't see the wires. The sequence is as follows:

1. Distribution lines carry the electricity in three phases over three wires from the distribution stations to your neighborhood.

2. When the power lines reach the neighborhood, the three phases are split up. Each of the three phases goes to a different block of houses in the same general neighborhood.

3. Very close to your home there is a pole with a single phase transformer. This transformer steps-down the voltage to its final use of 120 Volts.

4. The single phase transformer has three service wires (Fig. 9.13). All three wires carry the same phase. Two wires carry 120 Volts of electricity into your home. The third wire carries the used electricity from your home to the ground outside.

5. The wires in a home are first divided into branch circuits. Each branch then provides power for a series of individual outlets, light switches, or lights. Each branch provides power for 8-10 outlets or lights.

6. After the current is used for the desired purpose, a neutral wire carries the used current to the ground outside.

The Final Transformer and the Three Service Wires

The final transformer just outside your home is a "single phase transformer with three service wires" (Fig 9.13). All three wires carry the same phase. However each wire has a different purpose.

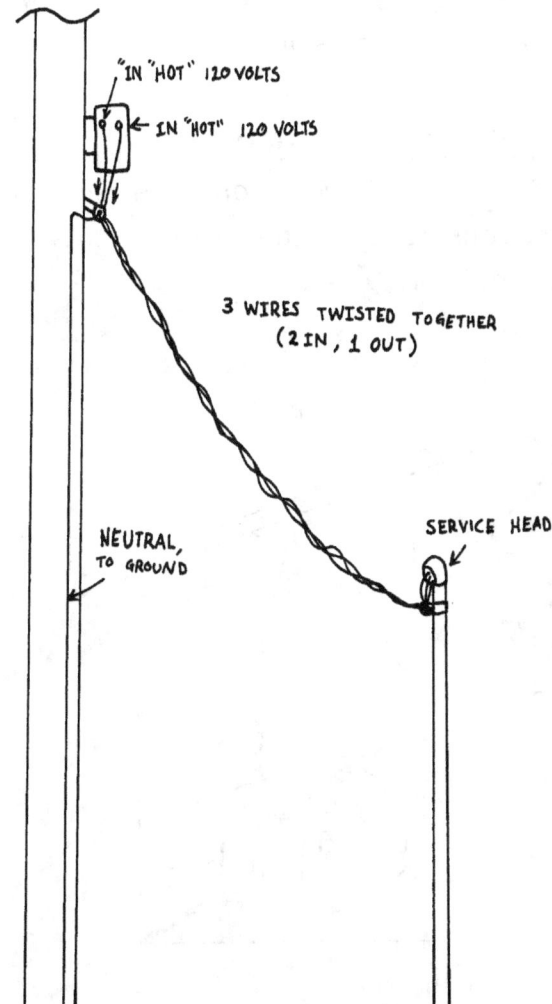

Hot Wires

Two of the wires carry electricity directly into your home. These are called the "hot" wires. Each hot wire carries 120 Volts.

Service Head

The hot wires enter a conduit through an opening called the "service head." The service head is curved in order to prevent rain from entering the conduit with the wires.

Neutral Wire

The third wire is neutral. After the electricity has been used it is sent out through this neutral wire. This wire leads back to the pole, then straight into the ground.

Fig. 9.13 Pole Transformer & Three Wires of Same Phase

These three wires are twisted together. Usually the neutral wire is the central wire, connected directly from your home to the pole. The two hot wires are then twisted around the neutral wire.

The hot wires are usually made of copper, for optimum conductivity. In contrast, the neutral wire is often made of steel, for strength, and is surrounded by aluminum, for corrosion resistance. Remember that once the electricity is used we don't care about power loss, therefore this steel–aluminum wire is conductive enough for the exiting current.

Sequence Through an Appliance

To better understand the path of electricity in your home, we will look at a typical appliance. For example, consider the sequence of events as the electricity travels through a toaster (Figure 9.14):

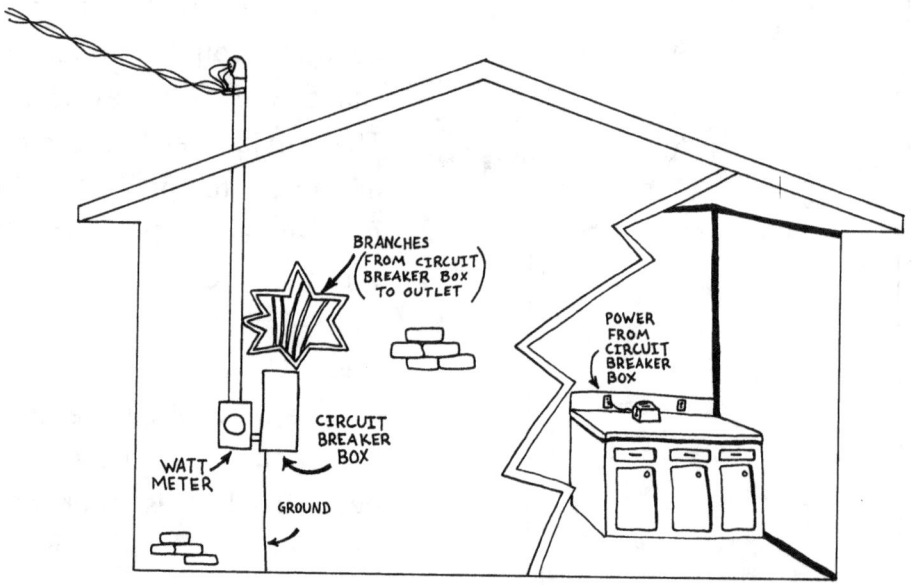

Fig. 9.14 Wiring in the Home

1. You turn the toaster on. This draws current from the power system.

2. 120 Volts of electricity goes to your home, through the watt meter outside, and into the circuit breaker box (or the fuse box).

3. From the circuit breaker box, the wires extend into several branches. Each branch delivers current to various rooms of the home.

4. The branch leading to your kitchen wall eventually connects to several outlets, including the outlet to your toaster. This branch also connects to the lights and switches in the kitchen. Each device is connected in series.

5. Plugging the power cord of your appliance into the wall allows the current to travel to the appliance. Current will flow into the device. However, no function is being performed until you turn the device on.

6. When you turn your toaster on, the wiring in the toaster becomes connected to the wiring in the wall. This allows the current to travel through the toaster to perform its task.

7. After the toaster has performed its task the current leaves the toaster through the exit wire.

8. The exiting current then travels through the neutral wire of the outlet. The exiting current travels through the house and to the outside, always through the neutral wire.

9. The used electricity is then conducted to the ground outside.

Neutral Wires and Ground Wires

Introduction

The electricity goes into the ground when it is done. Thus, the overall path of electric current starts in the power plant, then ends up in the earth near the homes.

When current leaves the appliances, the electricity travels through the series of exit wires in the home, eventually leading to the one neutral wire that runs outside the home. This final neutral wire leads to the pole by the street (where the transformer is located). From that pole the electricity is conducted straight down into the ground.

Note that in an underground cable system, the same procedure occurs, we just don't see the exiting neutral wire.

There are actually two ways for used electricity to exit a building and reach the ground: a neutral return wire or a ground wire. Both wires are similar, and the terms are often used interchangeably.

Similarities

Neutral wires and ground wires are similar in many ways:

a. Both neutral and ground wires carry the exiting current.

b. Both the neutral wire and the ground wire are neutral to begin with. The only current that flows through either of these wires is the current that leaves a device after we use it.

c. Both neutral and ground wires eventually go to ground. Both wires eventually lead to earth.

Differences

However, there are also some important distinctions between the neutral wire and the ground wire. Neutral wires are part of the normal path, whereas ground wires are used in emergencies.

The minimum wiring needed in residential wiring system is a hot wire (for incoming current) and the neutral return wire (for exiting current). However, the ground wire is often installed as a safety feature, providing a safe path for surges of electricity.

Under normal use no current flows through to the ground connection. Yet if there is a surge of extra current then the ground wire takes that extra current and channels that current safely to the earth.

The ground wires are usually connected to the outside of your water pipes. The pipes are usually metal and thus these pipes conduct the electricity effectively into the ground. Metal poles are also put into the ground for the same purpose.

120 Volts, 240 Volts, Prongs, Outlets

Introduction

Most devices in the home require only 120 Volts. However, some appliances require more power, which means using 240 Volts. Common 240 Volt devices include: stove, oven, air conditioner, water heater, and clothes dryer. All other appliances use 120 Volts.

The wiring must be installed specifically for each Voltage. The outlets are also designed differently for safety reasons. Because of these concepts, you cannot just plug a 240 Volt device anywhere in your home.

120 Volt (Figure 9.15)

Most appliances in the home use 120 Volts and are connected to the wall with 2 prongs. One prong carries the current into the appliance, the other prong carries the used current out of the appliance into the wall.

Fig. 9.15 The 120–Volt Outlet

Most 120 Volt appliances have cords with only two prongs: one prong for the current coming in, one prong for current going out. However, some devices have a third prong. The third prong is for safety only and is connected to ground. If there is a surge of electricity, then the ground wire will take that extra current and send it safely into the earth.

240 Volt (Fig 9.16)

A few major appliances, such as a stove, require 240 Volts. The wiring and outlet system are completely different for 240 Volts than for 120 Volts. In the 240 Volt system, the top two slots are both wired for incoming electricity. Behind the wall, there are *two* 120 Volt wires. Therefore, when you plug the stove into that wall, you have two wires, of 120 Volts each, entering your stove.

We can also now see the reason for the value of 240 Volts. If an appliance requires more voltage than 120 Volts, then two 120 volts wires enter the appliance at the same time. Two wires x 120 Volts each = 240 Volts.

The bottom slot is for the exiting current. After the electricity has been used in the appliance, the electricity is sent through the third prong of the cord, and into the third slot of the outlet. This bottom slot is connected to a ground wire which leads directly to the earth below the home.

Note that there are varieties of 240 Volt outlet and prong designs. The differences have to do with differing city regulations, or accommodating different amperage wires.

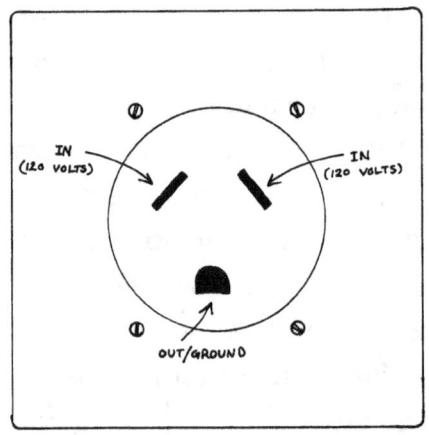

3-prong, 240-Volt Outlet

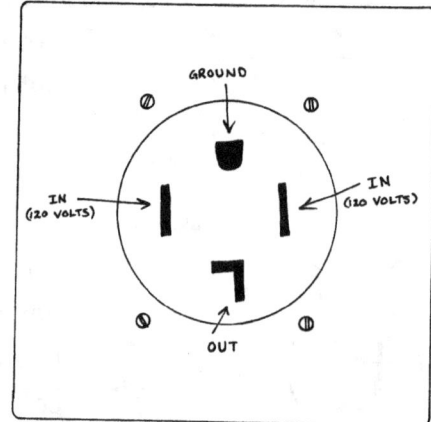

4-prong, 240-Volt Outlet

Fig. 9.16   240-Volt Outlets

Colors of Wires

The color of each wire is important. The coloring tells us clearly which wire performs which function. Common wire colors and their meanings are as follows:

1. Black: Incoming current, hot wire, 120 Volts
2. Red: Incoming current, hot wire, 120 Volts
3. White: Exiting current, neutral wire, leads to ground
4. Green (or other color): Exiting voltage, neutral wire, leads to ground

Fuses and Circuit Breakers

Introduction

When wires get too hot they can start a fire. Instead of creating an electrical fire it is best for the power to be disconnected. This is the purpose of fuses and circuit breakers. Fuses and circuit breakers will disconnect the power along particular branches before the faulty power line starts an electrical fire.

Operation of Fuses and Circuit Breakers

Each branch of wiring has sensors which measure an electrical factor (such as voltage, current, or temperature). If the value becomes too high then the sensor will trigger the fuse or circuit breaker. The power will immediately be prevented from traveling through that branch.

A fuse shuts the power off by melting. When a fuse gets too hot, the fuse melts, and the current is stopped through that wire. Note that a fuse is permanently destroyed. The fuse must be replaced before the wire can carry current again.

Circuit breakers last longer because they are not destroyed. The key component in a circuit breaker is a piece of metal that bends when heated. When a circuit breaker gets too much heat, the piece of metal bends, which then flips the circuit breaker switch. This action disconnects the wire. To reconnect the power, we flip the switch. This reconnects the wire, and also resets the metal component. Everything is then in place as before, and the circuit breaker will be able to perform its job again.

Safe Power Level and Safe Temperature Level

Each branch of the wiring in your home is designed to take a certain amount of power, voltage, and temperature. However, in order to be safe the fuses and circuit breakers disconnect the wire before any maximum is ever reached.

We calculate the "safe power level", which is usually 80% of maximum power. We can also calculate the "safe temperature level" which is usually 70%–80% of the maximum temperature. When the power or temperature reaches this safe level value, this factor becomes a trigger for the circuit breaker (or fuse) to disconnect the wire from the system.

To the Businesses

Introduction

The wiring in a business is similar to the wiring in homes. However, there are two distinct differences between wiring in the homes and wiring in the businesses: larger amounts of power are required, and there are multiple types of power needs.

1. Larger businesses, particularly manufacturing facilities, require large amounts of power to operate equipment. Furthermore, the equipment often requires larger voltages. There are several ways a business can increase its power: a) use larger wires, b) use multiple wires, or c) use higher starting voltages (by using a different final transformer).

2. Businesses also have multiple power needs. For example, a typical manufacturing facility will have air conditioning units, manufacturing equipment, and computers, each of which will each require different voltages. A hotel will have ovens and laundry equipment, yet also have lighting and televisions. Therefore, many larger businesses will need special combinations of wires, each wire carrying different voltages.

Two Phase and Three Phase Wiring

Most wiring in buildings, such as wiring in the home, is one phase. However, if there are numerous customers at one site then two phase and even three phase wiring may be used.

Recall what happens for most homes and most small businesses: electricity is carried over three wires in three phases from the power plant to the homes. As the electricity approaches the neighborhood, the phases separate. Each phase goes to a different block of homes. Now suppose that we have the same amount of customers in one building that we did in all those homes. In order to meet the needs of those customers we need more electricity. Therefore, rather than separating the phases to different buildings, we bring two of those phases or even three of those phases into that one building.

That concept is known as a "two phase" or a "three phase" wiring system. Buildings which use two or three phases include large office buildings, hotels, shopping centers, schools, and manufacturing facilities.

Multiple Power Level Needs for Businesses

Most businesses have multiple power level needs. Therefore a large business requires combination of wires, each wire carrying a different voltage. For example, a typical manufacturing facility requires several voltage lines: 120 volts for lights and computers; 240 volts for air conditioning; and 480 volts and higher for the manufacturing equipment.

Therefore, it is often necessary to have several final transformers, with each transformer reducing the voltage from the distribution line (typically 10,000 Volts) to different amounts (such as 120 or 240 Volts) before being sent into the building.

Therefore, by using a three-phase wiring system in conjunction with three different size transformers, the business can most effectively wire the building for its particular needs. For example, Phase 1 is reduced to 120 Volts, Phase 2 is reduced to 240 volts, and Phase 3 is reduced to 480 volts. Each power line is then sent to different regions of the facility. This is the most effective method for acquiring and delivering different voltage values in one facility, and it will be much easier to meet each of the power needs of the business.

Chapter Summary

1. The sequence of electricity from the local substation to the home is relatively straightforward.

   a. Distribution lines carry the electricity in three phases over three wires from the distribution stations to your neighborhood.

   b. When the power lines reach a neighborhood the three phases are split up. Each of the three phases goes to a different block of houses in the same general neighborhood.

   c. Near the home, there is a pole with a single phase transformer. This transformer steps down the voltage to its final use of 120 Volts.

   d. The single phase transformer has three service wires. All three wires carry the same phase. Two wires carry 120 Volts of electricity into the home. The third wire carries the used electricity from the home to the ground outside.

   e. The wires in a home are first divided into branch circuits. Each branch then provides power for a series of individual outlets, light switches, or lights.

2. Circuit breakers and fuses are designed to shut off a branch of current when the current reaches 80% of maximum power or temperature. This is done in order to prevent damage and potential fires.

3. Outlets in the home are wired for either 120 or 240 volts. For 120 volts there is one wire in, one wire out, and often a ground for safety. For 240 volts, there are two wires in, and one wire out.

4. After the current is used for the desired purpose a neutral wire carries the used current to the ground outside.

5. The wiring in businesses is similar to the wiring in homes, with two distinctions:
   a. Manufacturing facilities require large voltages in order to operate large equipment.
   b. Most businesses need a combination of wires, each wire carrying different voltages.

6. There are several ways a business can increase its power:
   a. Use larger wires
   b. Use multiple wires
   c. Use higher starting voltages, by using a different transformer

7. In a 2-Phase or 3-Phase Wiring System we bring all three phases into one building, rather than separate the phases into different buildings. These 2-Phase and 3-Phase Wiring Systems are used where there are numerous customers or numerous power requirements in one building.

8. The most effective method for a business to meet its various power needs is to use a 2-Phase or 3-Phase Wiring System, with each phase passing through a different size transformer.

9. Whether we have one phase, two phases, or three phases of wiring in a building, for each phase of electricity we will need:
   a. Hot wires carrying the electricity into the building
   b. Neutral wires carrying electricity out of the building
   c. Ground wires for safety

Conclusion

Many Americans hold passionate views about electrical power, yet few Americans understand all the details behind their passion. Electricity should not be mysterious. The science, the technology, and the data of electrical power can be understood by anyone.

Above all else, we must remember that there are no perfect solutions, there are only choices. Any option can be beneficial, yet each option has its own technical issues to work with. It is up to you and to your community to make those educated decisions. I hope that this book will help guide you in your choices.

M.F.

Appendix

Wire Sizes

Am. Wire Gauge #		Diameter, (inches)	Area (in²)	Area (cir. mils)
0000	(4/0)	.4600	.1662	211,600
000	(3/0)	.4096	.1318	167,800
00	(2/0)	.3648	.1045	133,100
0	(1/0)	.3249	.0829	105,600
1		.2893	.0657	83,690
2		.2576	.0521	66,360
3		.2294	.0413	52,620
4		.2943	.0328	41,740
5		.1819	.0259	33,090
6		.1620	.0206	26,240
7		.1443	.0163	20,820
8		.1285	.0123	16,510
9		.1144	.0103	13,090
10		.1019	.0081	10,380
11		.0907	.0065	8,230
12		.0808	.0051	6,530
13		.0720	.0040	5,180
14		.0641	.0032	4,110
15		.0571	.0025	3,260

Resistance in Wires

Resistance in wires is based on material of the wire and size of the wire. The resistances for the two most common materials are shown below. Wire size is measured in inches. Resistance is measured in Ohms per 1000 ft.

Am. Wire Gauge #		Diameter	Resistance Copper	Aluminum
0000	(4/0)	.4600	0.049	0.080
000	(3/0)	.4096	0.062	0.101
00	(2/0)	.3648	0.078	0.128
0	(1/0)	.3249	0.098	0.161
1		.2893	0.124	0.203
2		.2576	0.156	0.256
3		.2294	0.197	0.323
4		.2943	0.249	0.407
5		.1819	0.316	0.514
6		.1620	0.395	0.648
7		.1443	0.498	0.816
8		.1285	0.628	1.03
9		.1144	0.792	1.29
10		.1019	0.999	1.64
11		.0907	1.26	2.07
12		.0808	1.59	2.6
13		.0720	2.01	3.28
14		.0641	2.52	4.14
15		.0571	3.18	5.21
16		.0508	4.02	6.59

Remember that when conductors are stranded, the total resistance is based on the individual wires. Consider 19 Aluminum wires size gauge #10, stranded together: 19 conductors x 1.64 Ohms = 31.2 Ohms of resistance per 1000 feet.

Bibliography

<u>Electrical Principles</u>
(Generators, Turbines, Transmission Lines, Home Wiring)

1. <u>Electrical Power: Motors, Controls, Generators, Transformers</u>, by Joe Kaiser. Publisher: The Goodheart-Willcox Company, Inc.

2. <u>Electricity: Power Generation And Delivery, Sixth Edition</u>, by Alerick and Keljik, 1996. Publisher: Delmar

3. <u>Electricity: Motors, Controls, Alternators, Fifth Edition</u>, by Alerick and Keljik, 1991. Publisher: Delmar

4. <u>The Making of the Electrical Age</u>, by Harold Sharlin, 1963. Abelard-Schuman

5. <u>The Fantastic Inventions of Nikola Tesla</u>, by Nikola Tesla and David Childress, 1993. Publisher: Adventures Unlimited

6. <u>Electrician's Exam Preparation Guide</u>, by John Traister, 2001. Craftsman Book Company

7. <u>The Silent Energy</u>, by Kogan and Pick, part of the "Foundations of Science Library," 1966. Publisher: Greystone Press.

8. <u>Networks of Power: Electrification in Western Society</u>, 1880-1930, by Thomas Hughes, 1983. Publisher: Johns Hopkins University Press.

9. <u>The Lineman's and Cableman's Handbook, Ninth Edition</u>, by Kurtz, Shoemaker, and Mack, 1998. McGraw-Hill

10. <u>Wiring Essentials</u>, by various contributors, Black and Decker Quick Steps Series, 1996. Creative Publishing International.

11. <u>Home Electrical Wiring Made Easy</u>, by Robert Wood, 1993. Publisher: Tab Books, a division of McGraw-Hill

12. <u>Basic Home Wiring</u>, by various contributors, 1989. Publisher: Sunset Books

13. <u>Basic Wiring</u>, by various contributors, 1994. Publisher: Time-Life Books

14. <u>Advanced Wiring</u>, by various contributors, 1998. Publisher: Time-Life Books

15. <u>EMF and Power Lines - Report for the public</u>: *"Electric and Magnetic Fields Associated with the Use of Electric Power"*, EMF RAPID, of National Institute of Environmental Health Sciences, 2002. www.niehs.nih.gov/emfrapid/booklet/

Government Sites – General

1. US Department of Energy (DOE) www.energy.gov
2. US Department of the Interior www.doi.gov
3. Environmental Protection Agency (EPA) www.epa.gov
4. Food and Drug Administration (FDA) www.cfsan.fda.gov
5. National Institute for Occupational Safety and Health (NIOSH)
   www.cdc.gov/niosh
6. Federal Energy Regulatory Commission (FERC) www.ferc.gov

Department of Energy (DOE) Related Sites

1. Department of Energy (DOE) www.energy.gov
2. Energy Information Administration (EIA) www.eia.doe.gov
3. [Office of] Efficiency and Renewable Energy (EERE) www.eere.energy.gov
4. Office of Fossil Energy (in Dept of Energy) www.fossil.energy.gov
5. Electric Transmission and Distribution Office www.electricity.doe.gov
6. Science (Office of Science) www.sc.doe.gov
7. Nuclear Regulatory Commission (NRC) www.nrc.gov
8. Civilian Radioactive Waste Management (OCRWM) www.ocrwm.doe.gov
9. Yucca Mountain Project www.ocrwm.doe.gov/ymp/about/index.shtml
10. International Nuclear Safety Program http://insp.pnl.gov
11. International Nuclear Safety Center, Argonne Laboratory www.insc.anl.gov
12. National Energy Technology Laboratory (NETL) www.netl.doe.gov
13. National Renewable Energy Laboratory (NREL) www.nrel.gov
14. Oak Ridge National Laboratory www.ornl.gov
15. Los Alamos National Laboratory (LANL) www.lanl.gov/worldview
16. Pacific Northwest National Laboratory (PNL) www.pnl.gov
17. Starlight, from PNNL/DOE http://starlight.pnl.gov

List of Figures

Index for
Transmission of Electrical Power